AF802836

Keller's

DRIVER TRAINING MANUAL:

A Practical Guide To Tractor-Trailer Operations

© Copyright 1993

J. J. Keller & Associates, Inc.

3003 W. Breezewood Lane – P.O. Box 368
Neenah, Wisconsin 54957-0368
Phone: (414) 722-2848

United States laws and Federal regulations published as promulgated are in public domain. However, their compilation and arrangement along with other materals in this publication are subject to the copyright notice.

ISBN 1-877798-25-8

All rights reserved. Neither the Guide nor any part thereof may be reproduced in any manner without written permission of the Publisher.

First printing July, 1993.
Second printing December, 1993.

Due to the constantly changing nature of government regulations and other information, it is impossible to guarantee absolute accuracy of the material contained herein. The Publisher and Editors, therefore, cannot assume any responsibility for omissions, errors, misprinting, or ambiguity contained within this publication and shall not be held liable in any degree for any loss or injury caused by such omission, error, misprinting or ambiguity presented in this publication.

This publication is designed to provide reasonably accurate and authoritative information in regard to the subject matter covered. It is sold with the understanding that the publisher is not engaged in rendering legal, accounting, or other professional service. If legal advice or other expert assistance is required, the services of a competent professional person should be sought.

DRIVER TRAINING

Published & Printed by

J. J. Keller & Associates, Inc.

3003 W. Breezewood Lane – P.O. Box 368
Neenah, Wisconsin 54957-0368
Phone: (414) 722-2848

Publisher, John J. Keller

Compiled and Edited by R. Chris Halla

President, Robert L. Keller

Executive Vice President, James J. Keller

Vice President, Publications & Products, Terence J. Quirk

Research & Development Manager, Robert J. Fox

Publications Development Manager, David M. Wilhelms

Designer, Randy S. Heuer

The publisher wishes to thank the many instructors, safety directors and drivers who provided information and comment during the development of this manual. Special thanks are extended to those instructors who were kind enough to review the manuscript prior to production.

DIESEL INSTITUTE OF AMERICA
CDL
PRE TRIP SCORE SHEET

STUDENT'S NAME ______________________________ DATE ______________________

FRONT OF TRACTOR

1.	EXTERIOR CONDITION	YES____	NO___
	LEAKS, OIL, WATER, UNDER CAB	YES____	NO___
	WINDSHIELD	YES____	NO___
	WIPERS AND ARMS	YES____	NO___
	LIGHTS	YES____	NO___
	TURN SIGNALS	YES____	NO___

UNDER CAB

2.	OIL AND WATER	YES____	NO___
	WATER PUMP	YES____	NO___
	HOSES	YES____	NO___
	POWER STEERING FLUID	YES____	NO___
	BELTS	YES____	NO___
	ALTERNATOR AND BELT	YES____	NO___
	AIR COMPRESSOR	YES____	NO___
	FRAME	YES____	NO___

STEERING, SUSPENSION

3.	STEERING SHAFT AND U JOINT	YES____	NO___
	STEERING BOX	YES____	NO___
	PITMAN ARM	YES____	NO___
	DRAG LINK	YES____	NO___
	STEERING ARM	YES____	NO___
	TIE ROD END	YES____	NO___
	SPRING HANGERS	YES____	NO___
	SPRINGS	YES____	NO___
	U BOLTS	YES____	NO___
	SHOCK ABSORBER	YES____	NO___

FRONT WHEEL

4.	BRAKE PADS	YES____	NO___
	BRAKE DRUM	YES____	NO___
	SLACK ADJUSTER	YES____	NO___
	AIR HOSE	YES____	NO___
	BRAKE CHAMBER	YES____	NO___
	SEALS	YES____	NO___

RIM	YES____	NO___
TIRE	YES____	NO___
LUG NUTS	YES____	NO___
HUB OIL	YES____	NO___
AIR PRESSURE	YES____	NO___

DRIVER'S SIDE

5. DOOR AND HANDLE	YES____	NO___
WINDOWS	YES____	NO___
WINDOW CRANKS UP AND DOWN	YES____	NO___
MIRRORS AND BRACKETS	YES____	NO___
GRAB HANDLES	YES____	NO___
FIRE EXTINGUISHER & REFLECTORS	YES____	NO___
FUEL TANK, STRAPS, FUEL LEVEL	YES____	NO___
GASKET IN CAP, CROSSOVER	YES____	NO___

REAR OF TRACTOR

6. AIR AND ELECTRICAL CONNECTIONS	YES____	NO___
EXHAUST SYSTEM	YES____	NO___
REFLECTORS	YES____	NO___
CATWALK OR DECK PLATE	YES____	NO___
FRAME RAIL	YES____	NO___
DRIVE SHAFT AND U JOINTS	YES____	NO___
LIGHTS, STOP, TURN, TAIL	YES____	NO___

FRONT OF TRAILER

7. OVERALL CONDITION	YES____	NO___
AIR AND ELECTRICAL CONNECTIONS	YES____	NO___

FIFTH WHEEL

8. MOUNTING BOLTS	YES____	NO___
SAFETY LATCH	YES____	NO___
PLATFORM	YES____	NO___
RELEASE ARM	YES____	NO___
KING PIN/APRON	YES____	NO___
SLIDING FIFTH WHEEL	YES____	NO___
LOCKING PINS LOCKED	YES____	NO___

DRIVE WHEELS

9.	AIR HOSES	YES__/__	NO__/__
	BRAKE CHAMBERS	YES__/__	NO__/__
	SLACK ADJUSTERS	YES__/__	NO__/__
	BRAKE DRUMS	YES__/__	NO__/__
	SEALS	YES__/__	NO__/__
	BRAKE PADS	YES__/__	NO__/__
	RIMS	YES__/__	NO__/__
	SPACERS	YES__/__	NO__/__
	TIRES	YES__/__	NO__/__
	LUG NUTS AND WEDGES	YES__/__	NO__/__
	AXLE SHAFT AND NUTS	YES__/__	NO__/__
	AIR PRESSURE	YES__/__	NO__/__
	SPRING HANGERS	YES__/__	NO__/__
	SPRINGS	YES__/__	NO__/__
	U BOLTS	YES__/__	NO__/__

TRAILER SIDE

10.	MARKER LIGHTS	YES_____	NO_____
	REFLECTORS	YES_____	NO_____
	FRAME, CROSSMEMBERS, FLOOR	YES_____	NO_____
	LANDING GEAR	YES_____	NO_____
	DOOR TIES/HOOKS	YES_____	NO_____
	CRANK SECURED	YES_____	NO_____

TRAILER TANDEM

11.	AIR HOSES	YES__/__	NO__/__
	BRAKE CHAMBERS	YES__/__	NO__/__
	SLACK ADJUSTERS	YES__/__	NO__/__
	SEALS	YES__/__	NO__/__
	BRAKE DRUMS	YES__/__	NO__/__
	BRAKE PADS	YES__/__	NO__/__
	SPRINGS	YES__/__	NO__/__
	SPRING HANGERS	YES__/__	NO__/__
	U BOLTS	YES__/__	NO__/__
	RIMS	YES__/__	NO__/__
	TIRES	YES__/__	NO__/__
	SPACERS	YES__/__	NO__/__
	LUG NUTS AND WEDGES	YES__/__	NO__/__
	HUBS AND OIL LEVEL	YES__/__	NO__/__
	AIR PRESSURES	YES__/__	NO__/__

REAR OF TRAILER

12.	MARKER OR CLEARANCE LIGHTS	YES_____	NO_____
	TAILLIGHTS, STOP, TURN	YES_____	NO_____
	REFLECTORS	YES_____	NO_____
	LICENSE PLATE LIGHT	YES_____	NO_____
	REAR DOORS	YES_____	NO_____

DOOR HOOKS AND LATCHES	YES____	NO___
ICC BUMPER	YES____	NO___
MUD FLAPS	YES____	NO___

INSIDE OF CAB

13.	CLUTCH FREE PLAY	YES____	NO___
	GEARSHIFT FREE PLAY	YES____	NO___
	STEERING WHEEL PLAY	YES____	NO___
	AIR GAUGE	YES____	NO___
	OIL PRESSURE GAUGE	YES____	NO___
	WATER TEMP GAUGE	YES____	NO___
	BATTERY CHARGE GAUGE	YES____	NO___
	PARKING BRAKE	YES____	NO___
	SEAT BELT	YES____	NO___
	MIRRORS ADJUST BEFORE MOVING	YES____	NO___
	WINDSHIELD CLEAR	YES____	NO___
	WIPERS OPERATING	YES____	NO___
	WIPER FLUID	YES____	NO___
	HORNS BLOW	YES____	NO___
	LIGHT INDICATORS, HIGH BEAM, TURN, DASH LIGHTS	YES____	NO___
	HEATER/DEFROSTER	YES____	NO___

STATIC BRAKE CHECK

14.	AIR BUILD UP CHECK GOVENOR	YES____	NO___
	BRAKES RELEASE 3LB. AIR LOSS	YES____	NO___
	FOOT BRAKE APPLIED 4LB. AIR LOSS	YES____	NO___
	LOW AIR WARNING @ 60 PSI	YES____	NO___
	BRAKE BUTTONS POP OUT @ 25-45 PSI	YES____	NO___
	AIR GOVENOR OPERATING RANGE	YES____	NO___
	BRAKES ON MOVE VEHICLE	YES____	NO___
	HAND VALVE DOWN TRY TO MOVE	YES____	NO___
	PULL FORWARD 3-5 MPH. APPLY FOOT BRAKE	YES____	NO___

Form CDS-3 Meets U.S. Dept. of Transportation Requirements

DRIVER'S DAILY LOG

ORIGINAL - Submit to carrier
DUPLICATE - Retain in possession for 8 days

1	2	86
Month	Day	Year

Total mileage today:

Vehicle numbers - Show each unit: 200

I certify these entries are true and correct

Total miles driving today:

Driver's signature in full: Janine G. Pruett

Name of Carrier or Carriers: GRANGE TRUCKING

Name of co-driver: n/A

Main Office Address: WASHINGTON, D.C.

Home Terminal Address: BALTIMORE, MD.

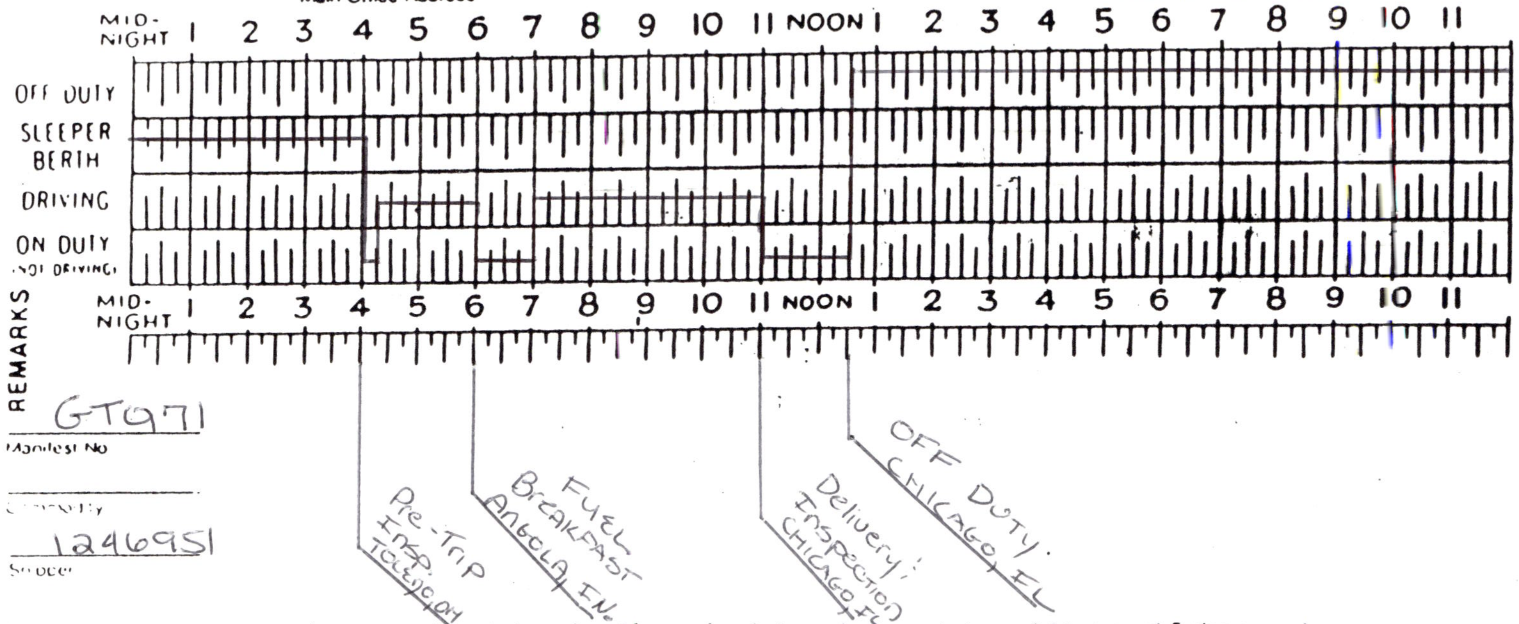

	Total Hours
Off Duty	11 1/2
Sleeper Berth	4
Driving	5 3/4
On Duty (Not Driving)	2 3/4
	24

Manifest No: GT971

Shipper: 1246951

Check the time and enter name of place you reported and where released from work and when and where each change of duty occurred. Explain excess hours.

FROM BALTIMORE, MD (Starting point or place) TO CHICAGO, IL. (Destination or turn around point or place)

USE TIME STANDARD AT HOME TERMINAL

RECAP
Eligible Hrs Today
Minus Hrs Today
Subtotal
Add Hrs Gain From First Day
Eligible Hrs Tomorrow

DRIVER'S DAILY LOG

ICC Form COS-3
Meets U.S. Dept. of Transportation Requirements

ORIGINAL - Submit to carrier
DUPLICATE - Retain in possession for 8 days

1 (Month) 1 (Day) 86 (Year)

Total mileage today: ____

Vehicle numbers - Show each unit: 200

I certify these entries are true and correct

Total miles driving today: ____

Driver's signature in full: [illegible]

Name of Carrier or Carriers: GRANGE TRUCKING

Name of co-driver: N/A

Main Office Address: WASHINGTON, D.C.

Home Terminal Address: BALTIMORE, MARYLAND

MID-NIGHT 1 2 3 4 5 6 7 8 9 10 11 NOON 1 2 3 4 5 6 7 8 9 10 11

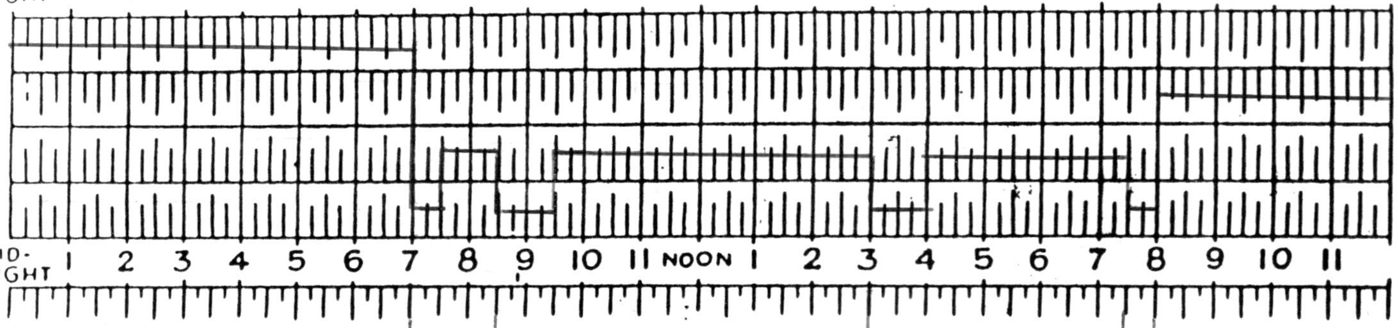

	Total Hours
OFF DUTY	7
SLEEPER BERTH	4
DRIVING	10
ON DUTY (NOT DRIVING)	3
	24

REMARKS

Manifest No: GT971

Shipper: 1246951

7: BALTIMORE MD. — Paperwork PreTrip I.
8–9: Breakfast Fuel FREDERICK MD.
3: LUNCH N. LIMA, OH
7–8: VEHICLE INS. TOLEDO, OH.
8: SLEEPER TOLEDO, OH

Check the time and enter name of place you reported and where released from work and when and where each change of duty occurred. Explain excess hours.

FROM BALTIMORE, MARYLAND (Starting point or place) TO CHICAGO, ILLINOIS (Destination or turn around point or place)

USE TIME STANDARD AT HOME TERMINAL

RECAP
Eligible Hrs Today
Minus Hrs Today
Subtotal
Add Hrs Gain From First Day
Eligible Hrs Tomorrow

DRIVER

MONTH/YEAR

BOOK-NUMBER

TERMINAL

8 DAYS—70 HOURS

	DAILY TOTAL DUTY & DRIVING HOURS	TOTAL ON DUTY HOURS LAST 8 DAYS	TOTAL ON DUTY HOURS LAST 7 DAYS	HOURS AVAILABLE FOR ON DUTY TIME NEXT DAY (70 MINUS COL. 4)
LAST SEVEN CONSECUTIVE DAYS IN PRECEEDING MONTH	0			
	0			
	10			
	10			
	10			
	0			
	0			
1	10	40	40	30
2	8	48	48	22
3	8	56	46	24
4	12	58	48	22
5	11	59	49	21
6	15	64	64	6
7	0	64	64	6
8	0	64	54	16
9	8	62	54	16
10	10	64	56	14
11	10	66	54	16
12	15	69	58	12
13	12	70	55	15
14	15	70	70	0
15	0	70	70	0
16	0	70	62	8
17	0	62	52	18
18	8 1/2	60 1/2	50 1/2	19 1/2
19	7 1/4	59 3/4	44 3/4	25 1/4
20				
21				
22				
23				
24				
25				
26				
27				
28				
29				
30				
31				

INSTRUCTIONS

This form is designed for use in maintaining a running record of a driver's on-duty time, regardless of the time your company designates as the start of its 24 hour period. It can also be used to accurately compute the driver's time used and hours available by making the entries in the appropriate chart for either a 70 hour-in-8 day or 60 hour-in-7 day work schedule.

← For those drivers working 70 hours in 8 consecutive days, use this chart

For those drivers working 60 hours in 7 consecutive days, use this chart →

Column 1 of each chart lists days of the month, 1–31 inclusive, plus space for the appropriate number of days from the preceding month that must be counted.

Column 2—In the proper chart, enter the total ON DUTY time (total of line 3—Driving time and line 4—On Duty, Not Driving, from the Driver's Record of Duty Status) for that day. If the driver did not work on any given day, enter a "0" for that day.

Column 3—Record, in the proper chart, the total number of hours accumulated in Column 2 for either the last 8 days (70 hour drivers) or 7 days (60 hour drivers). If over the allowable number of hours (60 or 70), there may be a violation.

Column 4—Record, in the proper chart, the total number of hours accumulated in Column 2 for either the last 7 days (70 hour drivers) or 6 days (60 hour drivers).

Column 5—Subtract the number entered in Column 4 from either 70 (70 hour drivers) or 60 (60 hour drivers) and enter on appropriate chart. This entry will indicate the number of on-duty hours the driver has available to him the next day before being in violation of either the 60 or 70 hour rule.

HOURS OF SERVICE CONTROL

The hours of service for each driver must be recorded on a separate sheet each month for purposes of control.

Violations of the 10, 15 and 60/70 hour rules are those of DRIVING after having attained the allowable number of hours driving or on-duty. These violations will not automatically show up in Column 3 and must be detected by reviewing the Driver's Record of Duty Status.

60 HOURS

	TOTAL DUTY & DRIVING HOURS	TOTAL ON DUTY HOURS LAST 7 DAYS	TOTAL ON DUTY HOURS LAST 6 DAYS	HOURS AVAILABLE FOR ON DUTY TIME NEXT DAY (60 MINUS COL. 4)
LAST SIX CONSECUTIVE DAYS IN PRECEEDING MONTH				
1				
2				
3				
4				
5				
6				
7				
8				
9				
10				
11				
12				
13				
14				
15				
16				
17				
18				
19				
20				
21				
22				
23				
24				
25				
26				
27				
28				
29				
30				
31				

Introduction

Driving a truck is a tough job. It isn't something just anyone can jump into the cab and do. It takes training. And it takes practice. This manual is intended to provide the basic information you need to learn to handle a combination vehicle. But the manual alone won't be enough. It will be most effective when you use it in a structured course led by an experienced trainer. (Although you may also refer to it in the future, as an experienced driver with a need to refresh your memory in certain areas.)

During the development of this manual, many instructors at both public and private schools, as well as trainers and safety directors within trucking companies, were asked what they needed in a training manual. They told us what was most important in a manual and how a manual should be structured to best fit the formats of their programs. They pointed out what worked (and what didn't) in the many training materials available. The result is a manual that we hope you and your instructor will find an indispensable part of helping you to become a well-trained driver.

In the course of your training, you may hear your instructor refer to the *Model Curriculum for Training Tractor-Trailer Drivers* and the *Model Driver's Manual for Commercial Vehicle Driver Licensing*. The *Model Curriculum* was developed as part of the government's Proposed Minimum Standards for Training Tractor-Trailer Drivers. The *Model Driver's Manual* was the model for most state Commercial Driver's License (CDL) manuals. Virtually all trainers and safety directors seem to agree that these publications are the basis for any good training program. For that reason, both contributed heavily to the form and content of this manual.

Additional material for this manual came directly from working truck driver trainers. They offered a wealth of information and methods they've used over the years to provide the best possible instruction to their students. Many safety directors, company executives and drivers themselves have also provided information to us over the years which we believe makes this manual an excellent training tool.

The sections and parts of this manual can be divided roughly into three areas: Those things you need to learn before operating a large vehicle; those things you need to learn to operate a large vehicle; and those additional things you need to learn to become a professional.

Throughout the manual, you'll find illustrations used to provide ready reference to the information being presented. Review questions following each major topic area will help you to determine whether or not you've understood the material. And at various points throughout the manual, under the heading "Experience Speaking," you'll have a chance to read what real drivers and others in the trucking business have to say on a number of specific topics. Plenty of space is provided throughout for you to write down your own notes, comments and questions for discussion.

Remember, this manual alone won't give you everything you need to learn to drive a big rig. It's a tool to be used by the student and instructor together. Learning to be a professional driver is a lifetime process. It begins with a good instructor, with a solid program and the information presented in this manual.

Table of Contents

Orientation 1

the trucking industry

a regulated industry

the professional driver

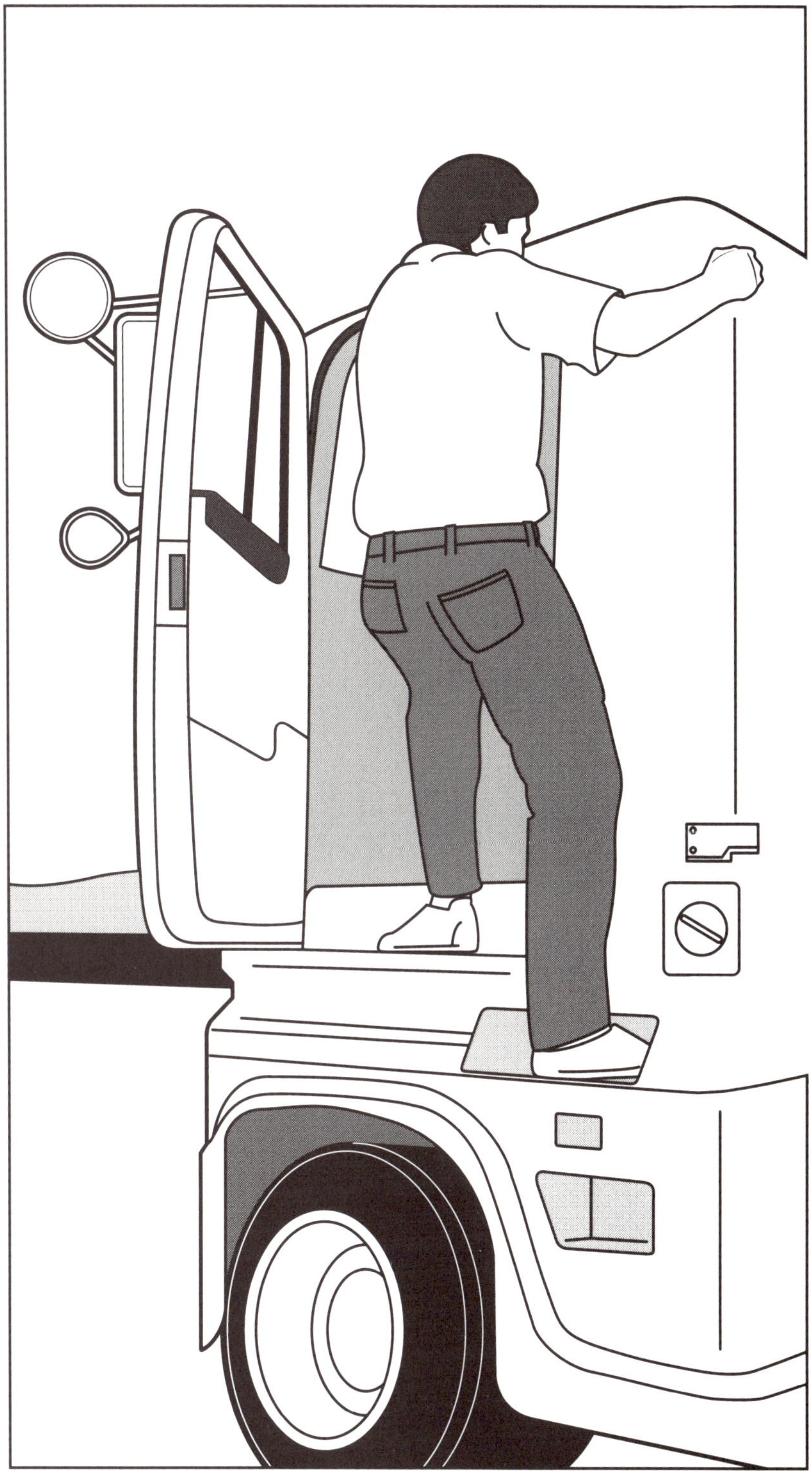

Orientation 1

The trucking industry is a vital part of our nation's economy. Other industries, commerce and the everyday lives of all of us rely on the trucking industry. If the trucks don't move, we all sit.

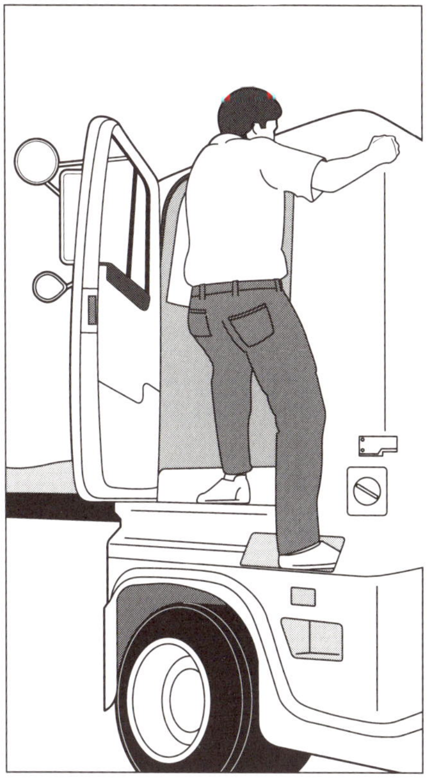

The Trucking Industry

Trucks haul agricultural products to processing plants, food and other goods to supermarkets and distribution centers. Raw materials such as iron, lumber and coal get to plants by truck. So does machinery. Without trucks, the fuel that goes to refineries, gas stations and our homes would remain unrefined in oil fields. Completed goods, from television sets to automobiles, would never reach retailers. Stores, homes and offices would largely be empty, if trucks didn't travel the highways.

Without the services of the trucking industry, our commerce might come to a grinding halt. There was a time in America before there were trucks, and before we needed trucks, but that time is long past. We have come to depend on the trucking industry in a way that almost defies description.

There are as many as four million tractor-trailer combinations on the road. Motor carriers transport more goods than rail, air or water. And even when other modes of transportation are used, the trucking industry often picks up where they leave off. Trucks haul three out of every four tons of goods.

A Regulated Industry

The trucking industry is subject to government regulations to — above all other things — ensure the safety of both the industry itself and the public. Part of being a professional driver is knowing and complying with regulations that affect the industry and its drivers. As a driver, you must make key regulations a part of your work habits.

There are two major types of trucking, intrastate and interstate. Intrastate hauling takes place within the borders of one state. Interstate hauling is done between states.

Interstate operations are regulated through the Federal Motor Carrier Safety Regulations (FMCSR). The FMCSR sets basic physical and knowledge requirements for drivers and establishes basic safety rules and standards for motor carriers, drivers and vehicles.

State regulations protect the citizens of the state and cause state records to be maintained. This includes driver licensing, road and fuel taxes, and length, width and weight limits.

The Professional Driver

The trucking industry employs as many as eight million persons in the United States. Among them, the driver is key. It's easy to see why there's always a need for qualified, professional drivers.

The professional driver meets or exceeds minimum standards set by government and the industry. The professional driver is in good physical and mental health at all times.

Professional drivers are skilled, safe, efficient, knowledgeable and fuel economy minded. They are good representatives of their employer and of the whole industry. A professional driver is one whom, in the eyes of the public, is expected to perform better in any given situation than the driver of a car.

orientation review

1. Name some products or materials hauled by truck.

2. Roughly how many tractor-trailer combinations are on the road right now?

3. Of every four tons of goods hauled, how much is hauled by truckers?

4. What is the purpose of government regulations?

5. What is the difference between intrastate and interstate transportation?

6. How many people in the United States are employed by the trucking industry?

7. What is a professional driver?

8. What does it take to become a professional driver?

Vehicle Controls 2

engine controls

primary vehicle controls

other primary controls

secondary vehicle controls

vehicle instruments

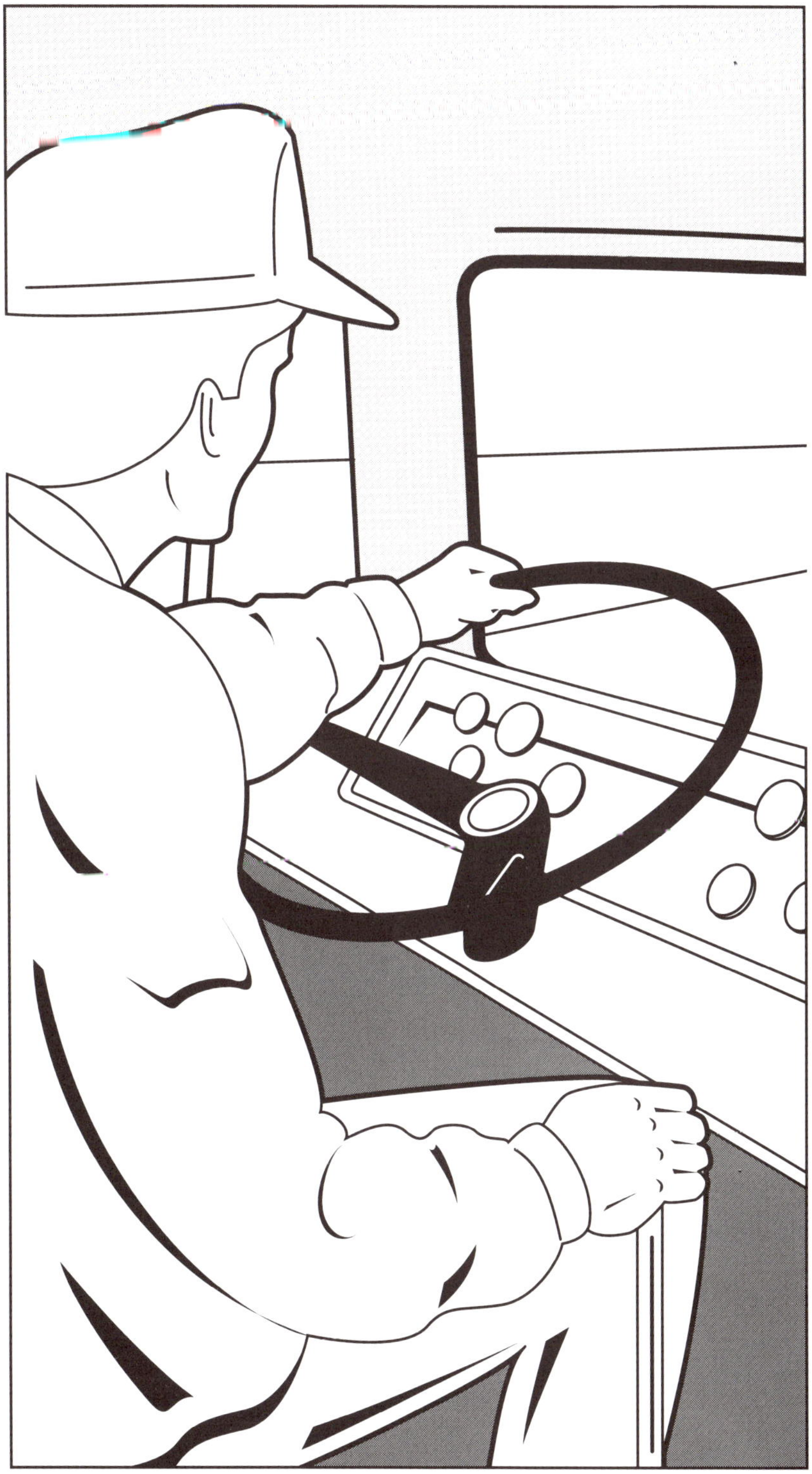

Vehicle Controls 2

When we talk about vehicle controls, we're talking about those in-cab devices that allow a driver to operate a vehicle.

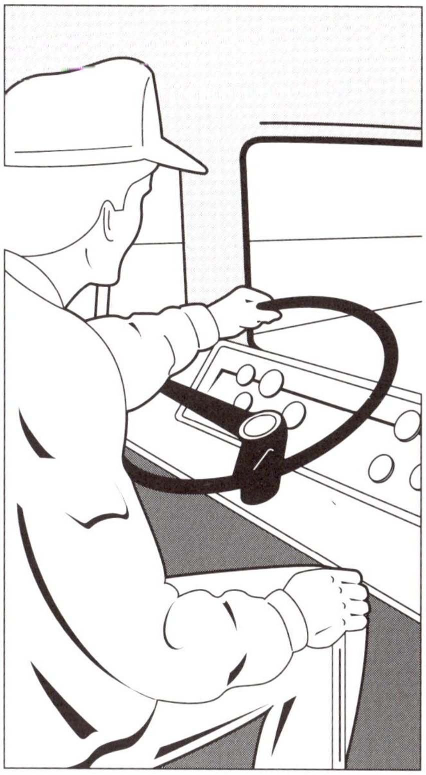

Large tractors, like automobiles, are built by many different manufacturers. Also like cars, while the shape and location of controls may vary somewhat, they perform essentially the same functions from vehicle to vehicle.

It's the driver's responsibility to learn the function of all controls before operating any vehicle. Never start out in a vehicle until you're familiar with that vehicle's controls.

There are three types of controls on a tractor-trailer: Engine controls; primary vehicle controls; and secondary controls. Engine controls are required to start and shut down a vehicle, as well as keep it running. Primary controls, such as the accelerator pedal, steering wheel, transmission controls and brake controls move the vehicle. Secondary controls aren't related directly to vehicle movement. These include light switches, climate controls and controls for starting and stopping the engine.

Engine Controls

Engine controls are similar in most vehicles. They may vary somewhat depending on the type of engine and engine manufacturer.

engine control switch

As on automobiles, the engine control switch energizes circuits and has three operating positions; on, off and accessory. The engine can't be started with the engine control switch off. With diesel engines, the engine control switch may incorporate an electric starting solenoid or electronic fuel cut-off switch.

hand throttle

The hand throttle, if the vehicle is so equipped, manually controls engine speed, or revolutions per minute (RPM). It's pulled out or upward to increase engine speed. You may also lock engine speed at a specific RPM level with the hand throttle. Electronic control or cruise control features fast idle.

starter button

The starter button is a momentary on-off switch located on the console. Pushing it in cranks the engine.

Primary Vehicle Controls

accelerator pedal

The accelerator pedal is operated the same on a truck as on an automobile. The accelerator pedal is depressed to increase speed and released to reduce speed.

steering wheel

Trucks have a larger diameter steering wheel than cars. This larger size gives the driver increased leverage to help overcome resistance. Steering a truck, naturally, requires more effort than steering a car.

It may take as many as eight full revolutions of a truck steering wheel to make a full right to left, or left to right, turn. Power and nonpower steering are both available. On some vehicles, the steering wheel can be adjusted up or down, in or out and tilted.

clutch pedal

The clutch pedal is used when cranking the engine and when shifting gears. There are two basic clutch positions, engaged and disengaged. When the clutch pedal is engaged, the driver's foot isn't on it, the engine and drivetrain are connected. The clutch pedal is disengaged when it's depressed, and the engine and drivetrain are separated. Freeplay is the amount of movement of the clutch pedal before the clutch actually starts to disengage. This generally isn't less than 1/2 inch or more than 2-1/2 inches. Extensive damage can be caused to the clutch or drivetrain if the clutch pedal isn't operated properly.

transmission controls

The exact type of transmission controls, or shifters, a vehicle has will depend on the type of transmission. Automatics don't use a clutch, just a gear selector lever. Manuals use a clutch and gear shift lever. Combination transmissions, such as main and auxiliary or main and variable speed rear-end units, have more controls to deal with.

air brake controls

The air brake control system is complex and involves a number of separate controls.

The foot brake control valve is also known as the foot treadle valve. Its function is to operate the service brakes in both tractor and trailer. By depressing it to open valves and send air pressure to the brake chambers, pressure is applied equally to all tractor and trailer service brakes. Foot brake pedals in cars create energy as the pedal is depressed; depressing a truck's brake pedal, when the truck has air brakes, releases stored energy. In use, the driver applies the greatest pedal pressure at first, then less as vehicle speed is reduced.

The trailer brake control valve is also called the hand valve, the trolley valve or the independent trailer brake. It operates the service brakes on the trailer only. And it's used only in special situations. The trailer brake control valve isn't used to hold equipment in place while parked or to stop with.

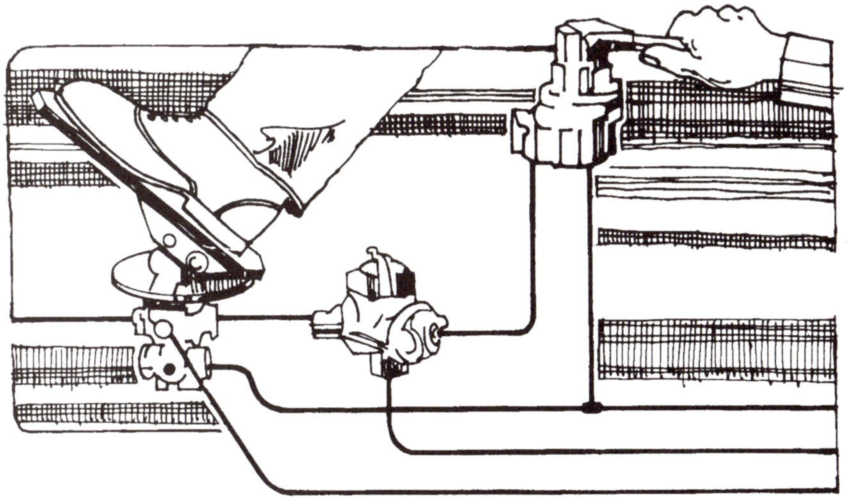

Most tractors use spring brakes operated by a dash mounted control called the parking brake control valve. The actual control may be a flip switch or a push-pull knob. The parking brake control valve should only be applied after the tractor is fully stopped.

emergency trailer braking controls

The tractor protection valve controls the compressed air supply to the trailer brakes only. It must be open (pushed in) for normal operation with a trailer and closed (pulled out) for operation on a tractor without a trailer.

In normal operation, the two-way switch located on the dash provides an air supply to the trailer brakes in the open position and closes off air to the trailer brakes in the closed position. The tractor protection valve closes automatically if the air supply drops to approximately 20-45 psi while driving. This protects the tractor's air supply by stopping the outflow of air. The emergency relay valve then opens and trailer brakes are applied. The tractor protection valve must never be used as a parking brake.

The trailer emergency relay valve functions only during an emergency related to lost air supply. It can't be counted on to hold brake pressure on a parked trailer.

Many trailers are equipped with spring loaded mechanical brakes on each axle. They're applied automatically when air pressure to the trailer is shut off or when the trailer is unhooked.

Other Primary Controls

In addition to the primary controls mentioned above, some vehicles are equipped with others. Some of them are of note here.

front brake limiting valve

The front brake limiting valve controls the amount of braking force on the front axle brakes. It will reduce braking force by approximately 50%, helping to prevent wheel lock-up on wet and slippery roads. The front brake limiting valve will be operated by a two-way switch on the instrument panel, or it may be automatic.

auxiliary brakes/speed retarders

Auxiliary brakes, or retarders, reduce the speed of a vehicle's forward motion independently of the service brakes. Their use reduces the burden placed on the primary brake system. They keep service brakes from being overused and thus overheated. There are four basic classifications of auxiliary brakes.

Exhaust brakes are considered the simplest form of heavy vehicle retarder. A butterfly valve is installed in the exhaust manifold that cuts off escaping exhaust. This builds up back pressure in the engine, preventing it from increasing speed. Exhaust brakes are usually operated by an on-off control switch in the cab or an automatic control switch on the accelerator or clutch.

Engine brakes are probably the most widely used type of retarder. They are built into the head of the engine and alter valve timing, turning the engine into an air compressor. They may be operated using a dash-mounted switch or automatically pre-selected to cut in when the foot is taken off the accelerator pedal.

Hydraulic retarders are one type of driveline retarder. They're usually mounted between the engine and the flywheel. During their use, oil is directed against vanes in a stator to slow the vehicle. They may be activated manually by a hand lever or accelerator switch, or they may be activated automatically.

Another type of driveline retarder is the electric retarder. It contains electro-magnets which exert a retarding force on rotors attached to the driveline. They're operated by a manual control switch in the cab.

interaxle differential lock

An interaxle differential lock locks and unlocks the rear tandem axles. In the unlocked position, axles turn independently of each other on a dry surface. In the locked position, power to the axles is equalized to help prevent wheels lacking traction from spinning. Interaxle differential locks should never be employed any longer than is absolutely necessary.

Secondary Vehicle Controls

There are a variety of secondary controls. Many are similar to controls with the same function in cars. Others are unique to heavy truck tractors. The number and kind of secondary controls will vary with vehicle design, as well as with options and equipment available.

Secondary vehicle controls may generally be classified within the following major categories.

Mechanical systems not related to driving.
Hydraulic cab tilt operation and fifth wheel lock controls

Seeing.
Light switches, windshield wipers and defroster.

Communication.
Horns, radio and lights.

Climate controls.
Heater and air conditioner.

Comfort controls.
Seat position control and air vents.

Driver Safety.
Seat belts.

Drivers must learn the function and operation of all secondary controls before operating a vehicle.

Vehicle Instruments

Vehicle instrumentation isn't complicated, but it's very important. Instruments monitor and report on the operating conditions of a vehicle and warn the driver of impending or immediate problems. It's the driver's responsibility to know when his vehicle has reached and when it's operating within a safe range.

Tachometer readings indicate when to shift. Temperature and pressure gauges warn of unsafe operating conditions. The speedometer monitors a vehicle's speed. Each instrument serves an important purpose. As a driver, you must be ever aware of what your vehicle's instrumentation is telling you. To do that, you have to understand your vehicle's instruments.

basic instruments

Speedometers are required by law. They show the speed of the vehicle in miles per hour (mph). Many drivers have found out the hard way that speedometers aren't always accurate.

The odometer indicates the number of miles travelled. The odometer can play an important role in on-the-road trip planning and verification, as well as providing data to both the driver and the company he works for.

The tachometer shows engine speed in hundreds of revolutions per minute (RPM). It serves as a guide for shifting to the appropriate transmission gear. What the appropriate RPM might be for any given engine speed varies according to engine and transmission make and model. You must learn the operating range for any vehicle you operate.

Fuel gauges indicate the amount of fuel remaining in the tank. Fuel gauge readings can only be considered a close estimate at best. You need to check tanks visually before each trip.

electrical system

The voltmeter measures the battery charging voltage. Normal readings or range will be specified in your operator's manual (usually 13-15.5).

The amount the battery is being charged or discharged is measured by the ammeter. Some vehicles will only have a warning light. Others may have gauges that say alternator instead of ammeter. Normal operating range is generally when the ammeter reads slightly to the charge side. Problems are indicated by continuous high charge or continuous discharge (the latter of which indicates the battery isn't receiving charge from the alternator).

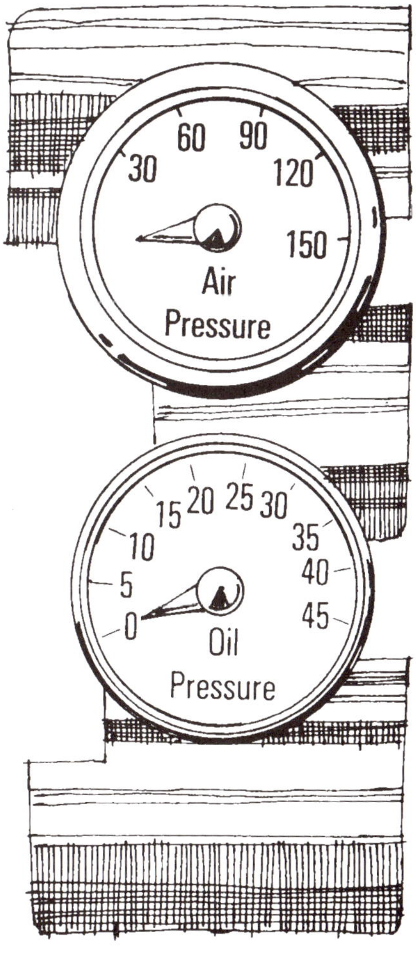

pressure gauges

The air pressure and oil pressure gauges are the two most important pressure gauges. Improper indications on either require immediate attention.

The air pressure gauge measures pressure in the reservoir, or tanks. A vehicle without sufficient air pressure shouldn't be operated. When the engine starts, air pressure starts building. It continues to build until maximum pressure (generally about

95-125 psi) is reached. The air pressure gauge will tell you if air pressure isn't building or if there's excessive loss when brakes are applied.

Oil pressure should register on the oil pressure gauge within a few seconds after the engine is started and gradually (but not too gradually) rise to normal operating range. The oil pressure gauge tells you if the engine is being lubricated properly.

Normal oil pressure operating range will depend on the specific vehicle and engine RPM. You should stop and investigate any time pressure doesn't register or when there's a gradual or sudden loss in pressure below normal operating range.

Vehicles can and often do have other pressure gauges. Some of them include coolant pressure, fuel pressure, clutch fluid pressure, manifold pressure and turbo pressure, in addition to others. Drivers should refer to the operator's manual for information on gauges, including proper and improper ranges and the consequence of operating under pressure.

coolant temperature gauge

The coolant temperature gauge measures the temperature of water and coolant in the engine block. Normally, engines should be operating at 170-195° F, or whatever is indicated as a safe range in the operator's manual. When engine and coolant are cold, the thermostat is closed, and fluids circulate through the engine only. The thermostat opens when the temperature reaches normal operating range, and coolant flows through the radiator. Overheating can be caused by a number of things:

Insufficient coolant;

Loose or broken fan belt;

Nonfunctioning thermostat, coolant pump or radiator shutters;

When engine is under a severe load condition.

Temperatures need to be monitored to avoid operating the engine at temperatures that are either too high or too low. (Remember, not letting an engine warm up properly can cause damage.) An overheating engine should be shut down at once.

The engine oil temperature gauge will normally show 180-225° F (although temperatures might run as high as 250-265° F with a full load, for a short period of time). As with all gauges, the operator's manual needs to be checked for normal ranges on each specific vehicle. As a general rule, oil temperature will run 20-60° F higher than water temperature. Oil temperature and oil pressure are related: High oil temperatures cause thinner oil and a decrease in pressure (normal oil pressure is about 40-60 pounds).

Some vehicles have additional temperature gauges. Always check the operator's manual.

warning devices

Vehicles are also equipped with warning lights and buzzers to let the driver know pressures and temperatures have reached a danger point. Some common warning devices are listed below. The operator's manual will provide information on warning devices and their operation and function.

The low air pressure warning alarm or light indicates inadequate pressure in the air brake system.

The coolant level alarm lights up when water level starts dropping, indicating a probable leak.

The oil level alarm lights up when the oil level becomes too low for normal operation.

The coolant temperature warning light goes on when the temperature becomes too high for normal operation.

The oil pressure warning light lights up when oil pressure becomes too low for safe operation.

The pyrometer warning goes on when exhaust temperatures are abnormal.

The differential warning flashes when the interaxle differential is in the locked position.

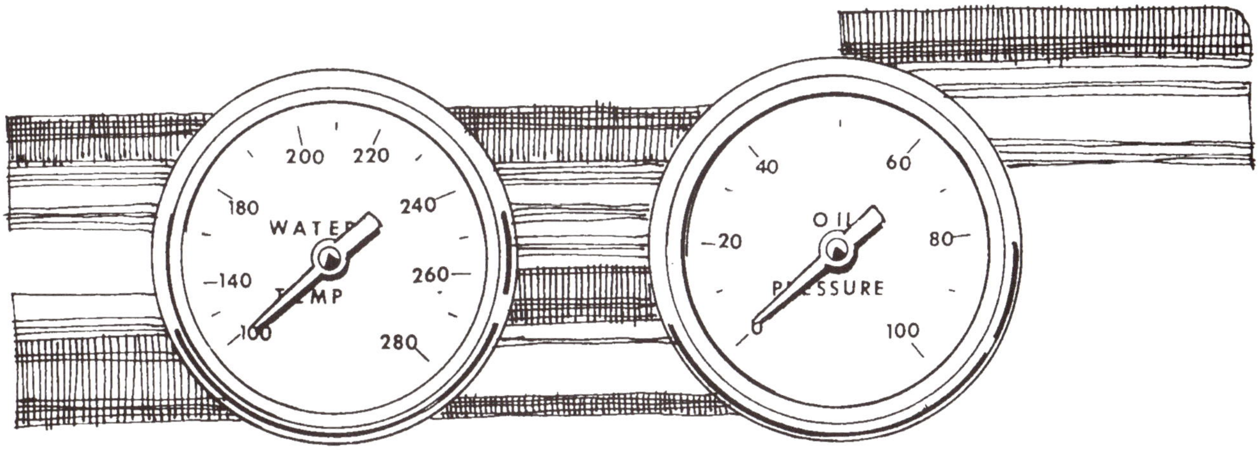

tachograph

The tachograph is one instrument the driver will usually have no control over. It is meant as a monitor for the motor carrier's use. Tachographs provide a record, in chart form, for every hour of a trip, whether the vehicle is parked and shut down or running. The information provided can show engine RPM, gear shifts and road speed.

Tachographs are seen as the enemy by some drivers, but they help motor carriers to analyze runs in both the areas of safety and economy. They can also provide essential information during accident investigations.

Obviously, what makes them unpopular with some drivers is that tachographs can be an excellent source of helping companies identify bad driving habits, such as long idles, over revving and erratic speeds.

More and more motor carriers are going to the use of on-board computers, which are capable of doing everything a tachograph can and much, much more.

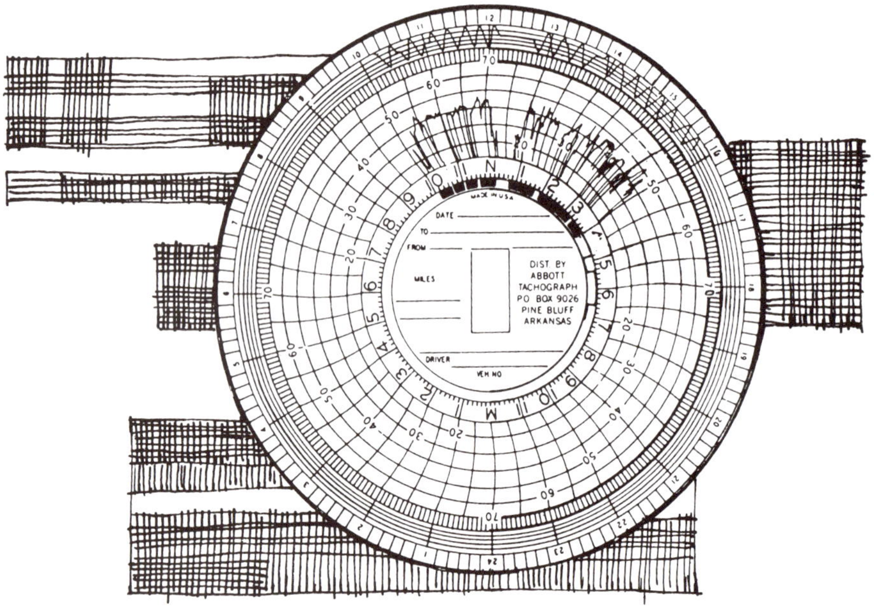

vehicle controls review

1. What do we mean when we talk about vehicle controls?

2. What are the three types of controls on a tractor-trailer combination?

3. Identify the basic engine controls.

4. Identify the primary vehicle controls.

5. Identify the secondary vehicle controls.

6. What is the purpose of auxiliary brakes and speed retarders?

7. Why is it important for you to be able to read and understand vehicle instruments?

Vehicle Systems 3

vehicle definition, cab design & axle arrangements

tractor construction

semi-trailers & full trailers

engine auxiliary systems

air braking system

Vehicle Systems 3

The professional driver's knowledge base begins with an understanding of the construction and systems of the vehicles being operated.

When the average person says "truck," the vehicle in question may be anything from a pick-up truck or the family van, to a delivery truck, all the way up to and including the largest combination vehicles (the latter being the vehicles we are most concerned with in this manual). Within the industry, we make clear there are many kinds of "truck", each with different driving demands. In later sections of this manual, when we say "truck," we'll be talking about truck tractors and combination vehicles.

Vehicle Definition, Cab Design & Axle Arrangements

straight truck & truck tractor

Straight trucks have the cab and cargo body on a single chassis. Truck tractors, on the other hand, are designed to pull other vehicles, or trailers, which carry cargo. Often we'll talk about the tractor plus the trailer as "the vehicle."

The chassis includes, all parts except the body and cab. The cab is the driver's compartment, control center and sometimes engine covering. The body (also known as the box or trailer) carries the cargo.

cab design

Most truck tractors fall into two categories: Conventional and cab-over-engine (COE). On a conventional, the cab sits behind the engine like on a car. The engine is gotten at by raising the hood. With a COE unit, the cab placement is over the engine, as the name implies. On a COE, the engine is accessed by tilting the cab forward.

Sleeper berths are available in both COE and conventional cabs. In both cases, the sleeper berth is directly behind the driver and passenger seats, or cockpit.

There are a few other cab designs besides conventional and COE. As a rule, the location or design of the cab is obvious from the name: For example, low cab forward (driver sits ahead of engine) and halfcab (driver sits beside engine).

tractor axle characteristics

The front tractor axle is the steering axle. The rear tractor axles are the power, or drive, axles. Live axles have power transmitted to them while dead axles don't.

Tractors are generally described by the number of live axles. For example, a tractor with two rear axles, but only one of them a live axle will be referred to as a single axle tractor. A tandem (two) axle tractor has two live axles.

A tag axle tractor has a dead axle behind the live axle which can, on some vehicles, be lifted when a load isn't being pulled. If the forward axle is dead and the rear a live axle, it's called a pusher axle, because the rear axle pushes the forward axle along.

When both axles are live, the tractor has the best traction and can pull the heaviest loads.

Tractor Construction

Tractors are designed piece by piece for specific hauling purposes. The cab design and axle arrangement are the most recognizable tractor design features, but other key components are selected and designed for specific purposes, based on the job, the terrain, size and type of payload and size restrictions.

frame

The frame is the basic foundation of the chassis. Its key parts are the left and right side frame rails, and the cross members connecting the frame rails. The frame is connected directly or indirectly to all parts of a truck. Most noticeably, it provides engine mount, cab support, fuel tank support and suspension mounting.

engine

The engine is the tractor's power plant. It will most often be fueled by either diesel or gasoline. Diesel engines are the most common. Diesel engines are internal combustion engines, like gasoline engines, but unlike many gasoline engines, they use no carburetor or spark plugs.

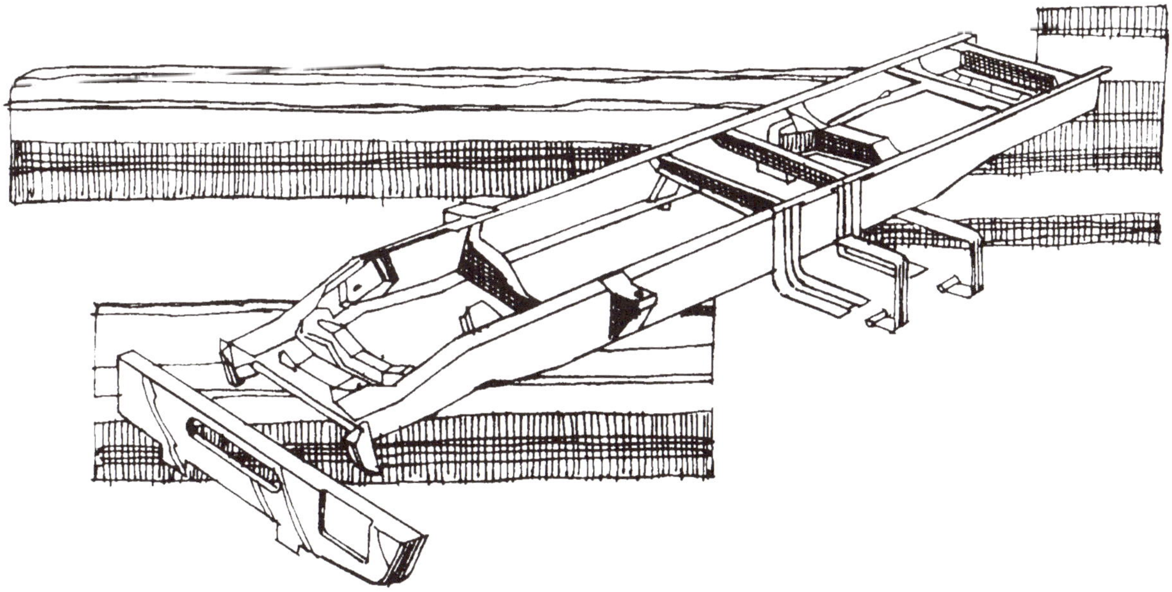

drivetrain

The drivetrain is made up, essentially, of the clutch, transmission, drive shaft and rear axle components.

The clutch is the driver's control that transmits power from the engine to the transmission.

The transmission, in turn, transmits power from the engine to the drive axle(s), using gears to increase or decrease speed. Transmissions may be manual, semiautomatic or fully automatic.

The drive shaft transfers power from the transmission to the drive axle(s). Universal joints keep the shaft transmitting the power to the rear axle over uneven surfaces.

An interaxle differential is used with tandem axles to divide power between axles and allow each axle to turn at a different speed. In the rear axle assembly, final drive gears operate the axle, and the axle differential allows wheels at each end of the axle to turn at a different speed when cornering. Multispeed ratio rear axles allow for two or three speeds in the final drive gears and are manually operated by the driver according to need.

suspension system

The suspension system supports the body and frame of the truck. It consists of the springs and the mounting brackets or hangers. The front or rear axles are attached to the suspension system on top of which rests the frame. With this arrangement, the axles can move up and down in response to ground changes without seriously affecting a load. By securing the suspension system at points spread throughout the frame, the stress of absorbing shocks can be well-distributed. The weight and durability of the suspension system, obviously, must increase if the truck is quite large and the cargo heavy. If the truck is smaller, the system doesn't have to be as sturdy.

The following characteristics are necessary for a good suspension system: The capacity to support the load; the ability to transmit full brake and tractive effort to the chassis frame; a cushioned ride for driver and cargo, laden or unladen; secure axles to assure correct driveline alignment; easy maintenance and lightweight parts.

steering system

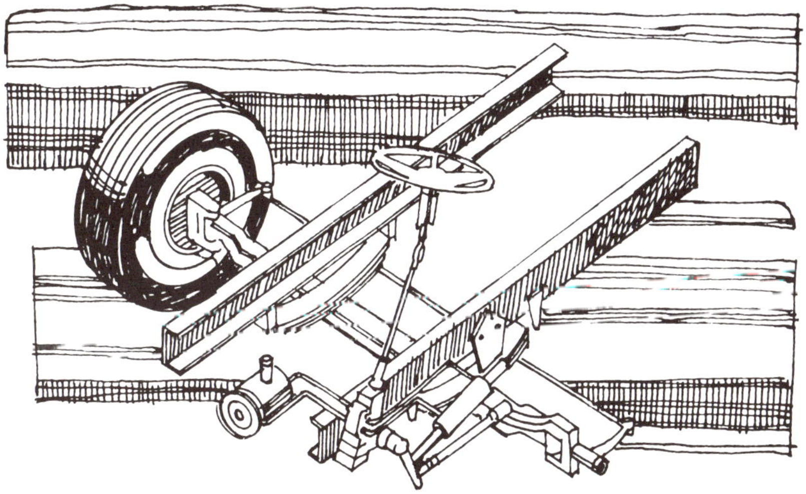

The principles of steering a truck tractor are basically the same as a car, but on a larger scale. The steering system is attached to the frame by the suspension system. The front axle is the steering axle. Key parts of the steering system are the steering knuckle, spindle, pitman arm and drag link, and steering gear box.

Steering knuckles are located at each end of the front axle, allowing the wheels to turn side to side. A spindle is at the end of each steering knuckle. Spindles are attached to the wheel assemblies and help allow each wheel assembly to rotate.

The pitman arm and drag link provide the connection between the steering wheel and steering knuckles by transferring motion from the steering wheel to the front wheels. They are connected through the steering gear box.

wheels & tires

Wheels may be either two-piece cast spoke assemblies or one-piece disc wheels. Variations on both are available from a number of companies.

Both tubeless and tube type tires are used. One typical size is 11.00 x 22.5, but other sizes are used depending on load requirements, special manufacturers' specifications and carrier preference.

Tire mounting must be done with care by a professional. It's important to mount tires properly on wheel rims. It's important not to mix styles (for example, radials and non-radials) on an axle. And it's very important to match duals by circumference (outer size).

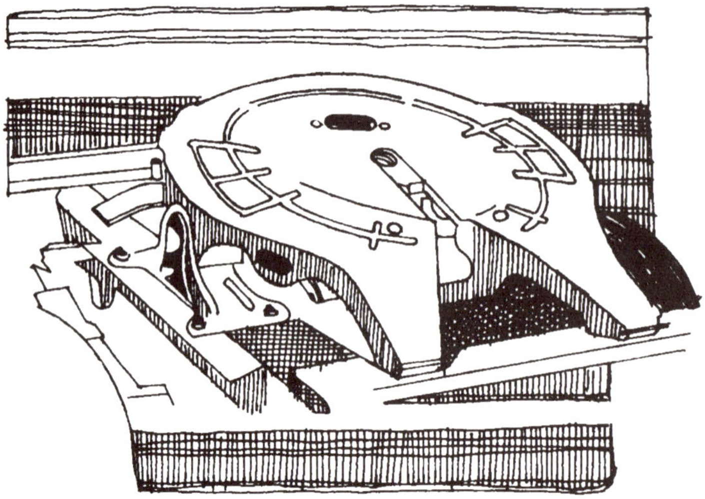

fifth wheel

The fifth wheel connects the trailer to the tractor. The lower half is on the tractor, the upper half on the trailer. The kingpin is part of the upper half. The kingpin and lower fifth wheel lock together.

Fifth wheels may be stationary or sliding. A stationary fifth wheel is bolted to the frame.

Manual sliding fifth wheels are mounted on a rail unit and move forward or backward. The axle weight distribution is locked in place by steel pins. A power sliding fifth wheel operates like the manual system using air pressure, with locks controlled from inside the cab.

Semi-Trailers & Full Trailers

Trailer design varies based entirely on the type and make-up (including weight) of the load to be carried.

Semi-trailers' rest on their own axles in the rear, while their front ends rest on the tractor. Full trailers have their own axles front and rear.

Trailers may be closed or open. Closed trailers include vans, refrigerated vans, tankers and open top vans. Open trailers might be flat beds, lowboys or dump trailers. Trailer length varies all the way from 18 feet to 60 feet.

semi-trailers

A semi-trailer is connected to a tractor by a king pin mounted in the trailer apron. The kingpin connects to the lower fifth wheel — on the tractor — allowing the tractor to turn under the trailer. The tractor's rear axle thus becomes the trailer's steering axle.

The trailer's landing gear, or dollies, are lowered when tractor and trailer are uncoupled to support the front of the trailer. The landing gear is raised when tractor and trailer are coupled.

Trailers may be equipped with one, two or more axles. Some axles may be retractable, some may be adjustable forward and backward.

full trailers

A drawbar is employed to connect a full trailer to a tractor. The eye of the drawbar attaches to a pintle hook. Safety chains are used in case the drawbar breaks.

A dolly, or converter gear, is sometimes used to convert a semi-trailer to a full trailer. The dolly, in such a case, will have spring suspension, drawbar and fifth wheel.

Engine Auxiliary Systems

A number of support systems are required for engine operation. Those engine auxiliary systems include the following:

Fuel system;

Air intake system;

Superchargers and turbochargers;

Exhaust system;

Cooling system;

Lubrication system;

Electrical system.

fuel system

The fuel system stores and provides fuel for engine operation. The fuel pump draws fuel from the fuel tanks through the fuel lines and fuel filter to the fuel injectors. The fuel injectors are operated by an injector pump or electronically and spray a fine mist of fuel into the engine cylinders.

air intake system

The air intake collects the huge amount of air needed for a diesel engine to function, filters out dust and dirt and distributes air to the cylinders via air intake ports.

superchargers & turbochargers

Superchargers and turbochargers force more air into engine cylinders than would otherwise occur. Engines without superchargers or turbochargers are known as naturally aspirated, because they inhale air normally. Superchargers are simply an air pump used to blow more air into the cylinders (which is why they're also known as blowers). Turbocharging is an alternative method of supercharging in which engine exhaust is used to power a turbine that blows more air into the cylinders. Superchargers and turbochargers not only create additional power, but they also contribute to better fuel economy and more complete burning of fuel (which results in less smoke).

exhaust system

The exhaust system collects engine exhaust in the exhaust manifold, muffles engine exhaust noise through a muffler and discharges exhaust gasses through a tailpipe or exhaust stack. The exhaust system is closed when the engine isn't running.

cooling system

The cooling system lowers the heat level of an operating engine. Coolant is stored in the radiator, circulated through engine water jackets by the water pump and returned to the radiator. Coolant temperature is controlled by the thermostat, and fan or radiator shutters. The thermostat regulates the flow of coolant through the radiator. The fan causes a flow of air through the radiator. Radiator shutters are controlled by the thermostat to control the amount of frontal air across the radiator.

lubrication system

The engine lubrication system provides clean lubricating oil so engine parts operate without friction. The oil supply is stored in the oil pan. The oil pump forces oil through the engine while filters keep the oil clean and free of debris. Several different types of filters and filtration systems are used. Diesels usually use both a primary and secondary filtering system. System also acts as a coolant, shock absorber and seal.

electrical system

The electrical system is a multipurpose system. The electrical system is used to crank the engine. It operates electrical equipment such as lights, instruments and radios, with batteries at its heart. On a combination unit, electrical current is supplied from the tractor to the trailer by a cord and plug. An alternator does most of the job. The battery is used for starting and overload situations.

Air Braking System

Perhaps more than any other vehicle system, the driver needs a complete understanding of air brakes. Here, more than any other system, you must be aware of the various parts of the system, how they work and how they relate to your responsibilities as a driver. If you haven't yet taken and passed your Commercial Driver's License (CDL) exam, you'll need this knowledge to acquire the license. A future employer will also expect you to have a working knowledge of air brakes.

Air brakes use compressed air to make the brakes work. You can apply all the braking force you need to each of the wheels of a heavy vehicle, even units pulling two or three trailers. Air brakes are a safe way of stopping large vehicles if the brakes are well maintained and used right. However, you must know more about air brakes than you need to know with the simpler brake systems used on light vehicles. Therefore, it's important

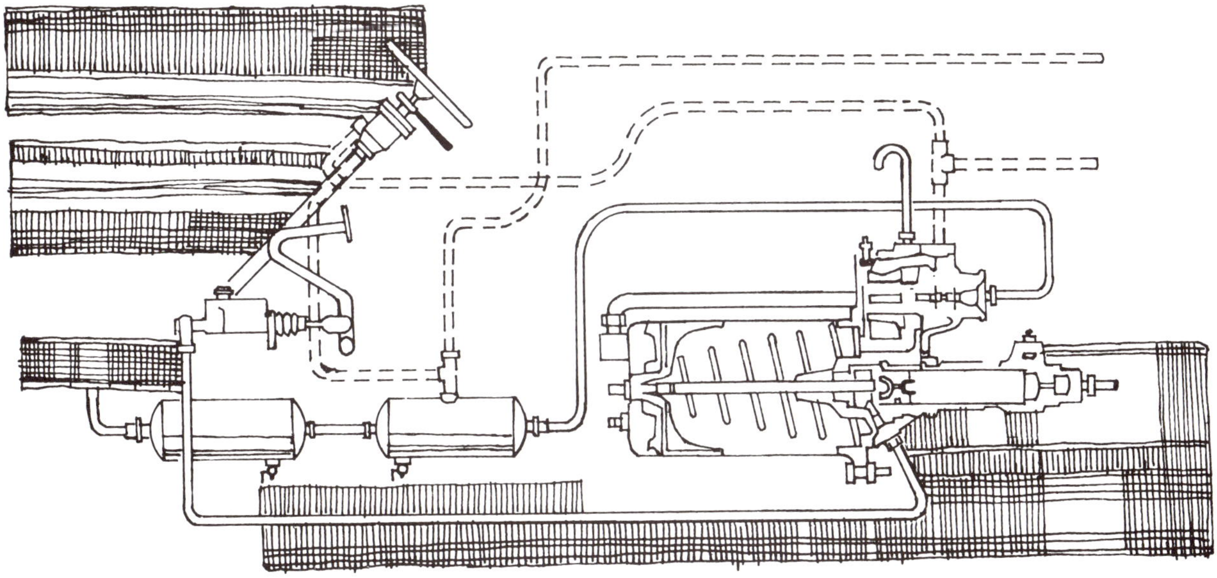

experience speaking

"The most common problem, in air systems, is having the brakes out of adjustment. Too long a push rod travel. The most common failures would be air leaks in the lines where they wear through, as well as ruptured diaphragms in the service brake and parking brake chambers themselves."

Francis
Maintenance Supervisor
California

for you to study and familiarize yourself.

Air brake systems are three braking systems combined according to law: The service brake system; the parking brake system; and the emergency brake system.

The service brake system applies and releases the brakes when you use the brake pedal during normal driving. The parking brake system applies and releases the parking brakes when you use the parking brake control on the dash. The emergency brake system uses parts of the service and parking brake systems to stop the vehicle in the event of a brake system failure.

parts of the air brake system

There are many parts to an air brake system. You should know about the parts discussed here.

The air compressor pumps air into the air storage tanks (reservoirs). The air compressor is connected to the engine through gears. The compressor may be air cooled or may be cooled by the engine cooling system.

The air compressor governor controls when the air compressor will pump into the air storage tanks. When air tank pressure rises to the cut-out level (around 125 pounds psi), the governor stops the compressor from pumping air. When the tank pressure falls to the cut-in pressure (around 100 psi) the governor allows the compressor to start pumping again.

Air storage tanks are used to hold compressed air. The number and size of air tanks varies among vehicles. The tanks will hold enough air to allow the brakes to be used several times, even if the compressor stops working.

Compressed air sometimes has some water and some compressor oil in it which is bad for the air brake system. For example, the water can freeze in cold weather and cause brake failure. The water and oil tend to collect in the bottom of the air tank and air valves. Therefore each air tank is equipped with a drain valve in the bottom. There are two types of drains.

Manually operated by turning a quarter turn or by pulling a cable.

You must drain the tanks yourself at the end of each day of driving.

Automatic - with which the water and oil is automatically expelled.

They may be equipped for manual draining as well.

The automatic types are available with electric heating devices. These help prevent freeze-up of the automatic drain in cold weather.

The alcohol container should be checked and filled, as necessary, every day during cold weather. Daily air tank drainage is still needed to get rid of water and oil (unless the system has automatic drain valves).

A safety relief valve is installed in the first tank the air compressor pumps air to. The safety valve protects the tank and the rest of the system from too much pressure. The valve is usually set to open at 150 psi. If the safety valve releases air, something's wrong.

You put on the brakes by pushing down the brake pedal (the foot valve, or treadle valve). Pushing the pedal down harder applies more air pressure. Letting up on the brake pedal reduces the air pressure and releases the brakes. Releasing the brakes lets some compressed air go out of the system, so the air pressure in the tanks is reduced. It must be made up by the air compressor. Pressing and releasing the pedal unnecessarily can let air out faster than the compressor can replace it. If the pressure gets too low, the brakes won't work.

When you push the brake pedal down, two forces push back against your foot. One force comes from a spring. The second force comes from the air pressure going to the brakes. This lets you feel how much air pressure is being applied to the brakes.

Foundation brakes are used at each wheel. The most common type is the S-cam drum brake.

Brake drums are located on each end of the vehicle's axles. The wheels are bolted to the drums. The braking mechanism is inside the drum. To stop, the brake shoes and linings are pushed against the inside of the drum. This causes friction which slows the vehicle (and creates heat). The heat a drum can take without damage depends on how hard and how long the brakes are used. Too much heat can make the brakes stop working. This is a constant consideration in downhill braking (to be discussed later).

When you push the brake pedal, air is let into each brake chamber. Air pressure pushes the rod out, moving the slack adjuster, thus twisting the brake cam shaft. This turns the S-cam (so called because it is shaped like the letter "S"). The S-cam forces the brake shoes away from each other and presses them against the inside of the brake drum. When you release the brake pedal, the S-cam rotates back and a spring pulls the brake shoes away from the drum, letting the wheels roll freely again.

With wedge brakes, the brake chamber push rod pushes a wedge directly between the ends of two brake shoes. This shoves them apart and against the inside of the brake drum. Wedge brakes may have a single brake chamber, or two brake chambers, pushing wedges in at both ends of the brake shoes. Wedge type brakes may be self-adjusting or may require manual adjustment. Those are most often found on construction equipment and off-road trucks.

In air-operated disc brakes, air pressure acts on a brake chamber and slack adjuster, like S-cam brakes. But instead of the S-cam, a power screw is used. The pressure of the brake chamber on the slack adjuster turns the power screw. The power screw clamps the disc or rotor between the brake lining pads of a caliper, similar to a large C-clamp.

Wedge brakes and disc brakes are less common than S-cam brakes.

All air-braked vehicles have a supply pressure gauge connected to the air tank. If the vehicle has a dual air brake system, there will be a gauge for each half of the system. (Or a single gauge with two needles). Dual systems will be discussed later. These gauges tell you how much pressure is in the air tanks.

The application pressure gauge shows how much air pressure you're applying to the brakes. (This gauge isn't on all vehicles.) When going down steep grades, increasing application pressure to hold the same speed means the brakes are fading. You should slow down and use a lower gear. The need for increased pressure can also be caused by brakes out of adjustment, air leaks, or mechanical problems.

A low air pressure warning signal is required on vehicles with air brakes. A warning signal you can see must come on before the air pressure in the tanks falls below 60 psi. (Or one-half the compressor governor cutout pressure on older vehicles). The warning is usually a red light or, a buzzer may also come on.

Drivers behind you must be warned when you put your brakes on. The air brake system does this with an electric switch that works by air pressure. The switch turns on the brake lights when you put on the air brakes.

Many vehicles have automatic front wheel limiting valves. They reduce the air to the front brakes except when the brakes are put on very hard (60 psi or more application pressure). These valves cannot be controlled by the driver.

All trucks, truck tractors, and buses must be equipped with emergency brakes and parking brakes. They must be held on by mechanical force (because air pressure can eventually leak away). Spring brakes are usually used to meet these needs. When driving, powerful springs are held back by air pressure. If air pressure is removed, the springs put on the brakes. A parking brake control in the cab allows the driver to let the air out of the spring brakes. This lets the springs put the brakes on. A leak in the air brake system which causes all the air to be lost will also cause the springs to put on the brakes.

Tractor and straight truck spring brakes will come fully on when air pressure drops to a range of 20 to 45 psi (typically 20 to 30 psi). Don't wait for the brakes to come on automatically. When the low air pressure warning light and buzzer first come on, bring the vehicle to a safe stop right away, while you can still control the brakes.

The braking power of spring brakes depends on the brakes being in adjustment. If the brakes aren't adjusted right, neither the regular brakes nor the emergency/parking brakes will work right.

In newer vehicles with air brakes, you put on the parking brakes using a diamond shaped, yellow, push-pull control knob. You pull the knob out to put the parking brakes (spring brakes) on, and push it in to release them. On older vehicles, the parking brakes may be controlled by a lever. Use the parking brakes whenever you park.

CAUTION. Never push the brake pedal down when the spring brakes are on. If you do, the brakes could be damaged by the combined forces of the springs and the air pressure. Many brake systems are designed so this won't happen, but not all systems are set up that way, and those that are may not always work.

It's much better to develop the habit of not pushing the brake pedal down when the spring brakes are on.

When main air pressure is lost, the spring brakes come on. Some vehicles have a separate air tank which can be used to release the spring brakes. This is so you can move the vehicle in an emergency. One of the valves is a push-pull type and is used to put on the spring brakes for parking. The other valve is spring loaded in the "out" position. When you push the control in, air from the separate air tank releases the spring brakes so you can move. When you release the button, the spring brakes come on again. There's only enough air in the separate tank to do this a few times. Therefore, plan carefully when moving. Otherwise, you may be stopped in a dangerous location when the separate air supply runs out.

The trailer hand valve (also called the trolley valve) works the trailer brakes. The trailer hand valve should be used only to test the trailer brakes. Don't use it in driving, because of the danger of making the trailer skid. The foot brake sends air to all the brakes on the vehicle (including the trailer). There's much less danger of causing a skid or jackknife when using just the foot brake.

Never use the hand valve for parking, because all the air might leak out, unlocking the brakes (in trailers that don't have spring brakes). Always use the parking brakes when parking. If the trailer doesn't have spring brakes, use wheel chocks to keep the trailer from moving.

The tractor protection valve keeps air in the tractor, or truck, should the trailer break away or develop a bad leak. The tractor protection valve is controlled by the trailer air supply control valve in the cab. The control valve allows you to open and shut the tractor protection valve. The tractor protection valve will close automatically if air pressure is low (in the range of 20 to 45 psi). When the tractor protection valve closes, it stops any air from going out of the tractor. It also lets the air out of the trailer emergency line. This causes the trailer emergency brakes to come on.

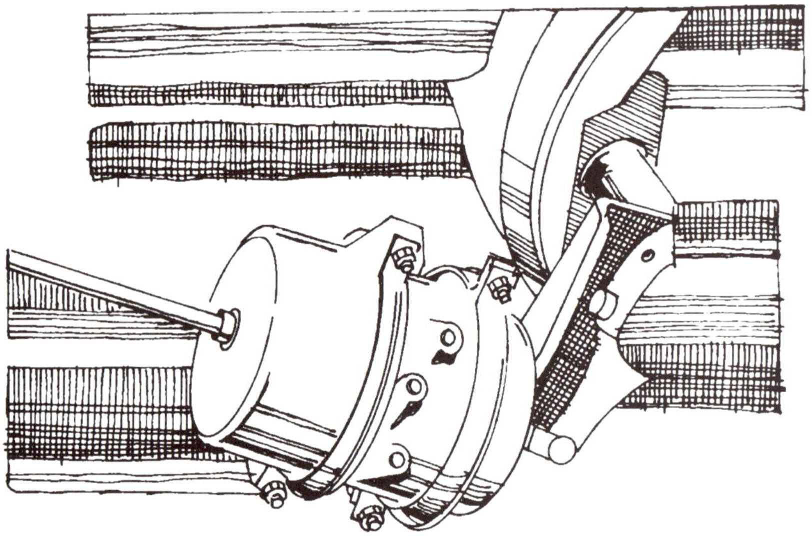

The trailer air supply control on newer vehicles is a red eight-sided knob used to control the tractor protection valve. You push it in to supply the trailer with air, and pull it out to shut the air off and put on the trailer emergency brakes. The valve will pop out (thus closing the tractor protection valve) when air pressure drops into the 20 to 45 psi range.

Every combination vehicle has two air lines, the service line and the emergency line. They run between each vehicle (tractor to trailer, trailer to dolly, dolly to second trailer, etc.)

The service line (also called the control line or signal line) carries air which is controlled by the foot brake or the trailer hand brake. Depending on how hard you press the foot brake or hand valve, the pressure in the service line will similarly change. The service line is connected to relay valve(s) on the trailer(s) to apply more or less pressure to the trailer brakes. The relay valve connects the trailer air tanks to the trailer air brakes. As pressure builds up in the service line, the relay valve opens and sends air pressure from the trailer air tank to the trailer brake chambers, thus putting on the trailer brakes.

The emergency line (also called the supply line) has two purposes. First, it supplies air to the trailer air tanks. Secondly, the emergency line controls the emergency brakes on combination vehicles. Loss of air pressure in the emergency line causes the trailer emergency brakes to come on. The pressure loss could be caused by a trailer breaking loose, thus tearing apart the emergency air hose. Or it could be caused by a hose, metal tubing, or other part which breaks, letting the air out. When the emergency line loses pressure, it also causes the tractor protection valve to close (the air supply knob will pop out).

Emergency lines are often coded with the color red (red hose, red couplers, or other parts) to keep from getting them mixed up with the blue service line.

Glad hands are coupling devices used to connect the service and emergency air lines from the truck or tractor to the trailer. The couplers have a rubber seal which prevents air from escaping. Clean the couplers and rubber seals before a connection is made. When connecting the glad hands, press the two seals together with the couplers at a 90 degree angle to each other. A turn of the glad hand attached to the hose will join and lock the couplers.

Some vehicles have dead end or dummy couplers to which the hoses may be attached when they're not in use. This prevents water and dirt from getting into the coupler and the air lines. Use the dummy couplers when the air lines are not connected to a trailer. If there are no dummy couplers, the glad hands can sometimes be locked together (depending on the couplings). It's very important to keep the air supply clean.

When coupling, make sure to couple the proper glad hands together. To help avoid mistakes, colors are sometimes used. Blue is used for the service lines and red for the emergency (supply) lines. Sometimes, metal tags are attached to the lines with the words "service" and "emergency" stamped on them.

If you do cross the air lines, supply air will be sent to the service line instead of going to charge the trailer air tanks. Air won't be available to release the trailer spring brakes (parking brakes). If the spring brakes don't release when you push the trailer air supply control, check the air line connections.

Older trailers don't have spring brakes. If the air supply in the trailer air tank has leaked away, there will be no emergency brakes, and the trailer wheels will turn freely. If you crossed the air lines, you could drive away but you wouldn't have trailer brakes. This would be very dangerous! Always test the trailer brakes before driving, with the hand valve or by pulling the air supply (tractor protection valve) control. Pull gently against them in a low gear to make sure the brakes work.

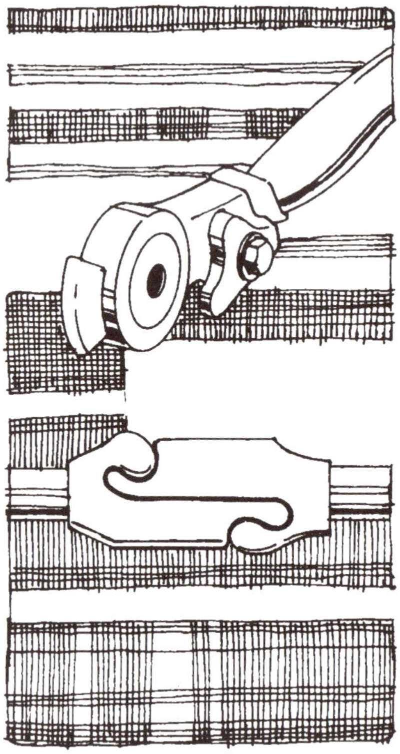

Each trailer and converter dolly has one or more air tanks. They're filled by the emergency (supply) line from the tractor. They provide the air pressure used to operate trailer brakes. Air pressure is sent from the air tanks to the brakes by relay valves. The pressure in the service line tells how much pressure the relay valves should send to the trailer brakes. The pressure in the service line is controlled by the brake pedal (and the trailer hand brake).

It's important that you don't let water and oil build up in the air tanks. If you do, the brakes may not work right. Each tank has a drain valve on it, and you should drain each tank every day. If your tanks have automatic drains, they will keep moisture out. But you should still open the drains to make sure.

Shut-off valves (also called cut-out cocks) are used in the service and supply air lines at the back of trailers used to tow other trailers. These valves permit closing the air lines off when another trailer is not being towed. You must check that all shut-off valves are in the open position except the ones at the back of the last trailer, which must be closed.

Newer trailers have spring brakes just like trucks and truck tractors. However, converter dollies and trailers built before 1975 aren't required to have spring brakes. Those that don't have spring brakes have emergency brakes which work from the air stored in the trailer air tank. The emergency brakes come on whenever air pressure in the emergency line is lost. These trailers have no parking brake. The emergency brakes come on whenever the air supply knob is pulled out or the trailer is disconnected. But the brakes will hold only as long as there's air pressure in the trailer air tank. Eventually, the air will leak away, and then there will be no brakes. Therefore, it's very important for safety that you use wheel chocks when you park trailers without spring brakes.

A major leak in the emergency line will cause the tractor protection valve to close and the trailer emergency brakes to come on.

You may not notice a major leak in the service line until you try to put the brakes on. Then, the air loss from the leak will lower the air tank pressure quickly. If it goes low enough, the trailer emergency brakes will come on.

dual air brake systems

Most newer heavy-duty vehicles use dual air brake systems for safety. A dual air brake system has two separate air brake systems which use a single set of brake controls. Each system has its own air tanks, hoses, lines, etc. One system typically operates the regular brakes on the front axle (and possible one rear axle). Both systems supply air to the trailer (if there is one). The first system is called the primary system. The other is called the secondary system.

Before driving a vehicle with a dual air system, allow time for the air compressor to build up a minimum of 100 psi pressure in both the primary and secondary systems. Watch the primary and secondary air pressure gauges (or needles, if the system has two needles in one gauge). Pay attention to the low air pressure warning light and buzzer. The warning light and buzzer should shut off when air pressure in both systems rises to a value set by the manufacturer. This valve must be greater than 60 psi.

The warning light and buzzer should come on before the air pressure drops below 60 psi in either system. If this happens while driving, you should stop right away and safely park the vehicle. If one air system is very low on pressure, either the front or the rear brakes will not be operating fully. This means it will take you longer to stop. Bring the vehicle to a safe stop, and have the air brake system fixed.

vehicle systems review

1. Why must air tanks be drained?

2. What is a supply pressure gauge used for?

3. What are spring brakes?

4. Do air brakes work instantly, like hydraulic brakes?

5. Why is it important to go slow on downgrades?

6. Using the brakes hard going downhill is okay if you let up on the pedal frequently to cool the brakes: True or False?

7. Why should you not use the trailer hand valve while driving?

8. Describer what the trailer air supply control does.

9. Describe what the service line is for.

10. What is the emergency air line for?

11. Why should you use chocks when parking a trailer without spring brakes?

12. From the driver's point of view, what is the most important vehicle system?

Vehicle Inspection 4

types of vehicle inspection

what to look for

a seven-step pretrip inspection method

on-the-road inspections

after-trip inspections

additional inspection procedures for special rigs

Vehicle Inspection 4

Safety is the reason why you learn how to do vehicle inspections before you ever drive a commercial vehicle.

Inspecting your vehicle helps you to know that it's safe. Furthermore, state and federal laws require vehicle inspection by the driver. State and federal inspectors also inspect commercial vehicles and will put an unsafe vehicle out of service until the driver or owner fixes it.

Types of Vehicle Inspection

Three separate kinds of inspection are required: Pretrip, on-the-road and post-trip, or after-trip.

pretrip

You do a pretrip inspection before each trip to identify problems that could cause a breakdown or accident.

on-the-road

For safety during a trip, you should watch your gauges for signs of trouble and use your senses (look, listen, smell, feel) to check for problems. Each time you stop, check all critical items: Tires, wheels and rims; brakes; lights; brake and electrical connections to trailer; trailer coupling devices; and cargo securement devices. Be aware that regulations have requirements for periodic stops to make safety inspections.

post-trip

You do a post-trip inspection at the end of a trip, day or tour of duty. The after-trip inspection should include filling out a vehicle condition report listing any problems found. The inspection report helps the owner and maintenance department know when something needs fixing and to make sure repairs are made.

What To Look For

tires

Look for tire problems. It's dangerous to drive with bad tires. Look for problems such as too much or too little air pressure and excessive wear. You need at least 4/32 inch of tread depth in every major groove on the front wheels. You need 2/32 inch on other wheels. No fabric should show through the tread or sidewall.

Other tire problems to check for include cuts or other damage, tread separation, duals that come into contact with each other or parts of the vehicle, mismatched sizes, mixing of radial and bias-ply tires, and cut or cracked valve stems.

While checking tires, also inspect wheels and rims. Damaged wheels or rims could cause an accident. A damaged rim can cause a tire to lose pressure or come off. Rust around wheel nuts may mean they're loose. After a tire's been changed, stop after driving a few miles, and recheck tightness of the wheel nuts. Missing clamps, spacers, studs or lugs mean danger. Mismatched, bent or cracked lock rings are also dangerous. Wheels or rims that have welding repairs aren't safe or legal.

brake system

Brake system air pressure checks are crucial. No audible leaks are permissible. Air pressure should be no more than three lbs/minute with the engine off and no brakes applied, and no more than four lbs/minute with the engine off and brakes fully applied (after initial application).

Also check for missing or defective gauges, and defective low air warning devices.

Brake drums need to be checked for cracks. Shoes or pads should be inspected for signs of oil, grease or brake fluid. Check for shoes that are worn dangerously thin (1/4 inch or less), are broken or missing. Brake chambers should not be missing or damaged. And always check slack adjusters.

Air brake lines need to be secured properly. They shouldn't show signs of hardening, swelling, excessive wear or damage. The air reservoir must be properly attached to the vehicle and shouldn't contain excessive water.

Brake lines to the trailer should not be tangled, restricted or damaged. Be certain that they are correctly attached and supported (not contacting catwalk, frame or other parts of the truck).

steering system

Steering system defects include missing parts, such as nuts, bolts and cotter keys, as well as bent, loose or broken parts, such as the steering column, steering gear box and tie rods. If your vehicle is equipped with power steering, inspect hoses, pumps and fluid level, and look for leaks. Steering wheel play of more than 10 degrees (approximately two inches movement at the rim of a 20-inch steering wheel) can make it hard to steer. Steering wheel play of 30 degrees or more is illegal.

suspension system

The suspension system holds up the vehicle and its load. It keeps the axles in place. Therefore, broken suspension parts can be very dangerous because a faulty suspension can, among other things, allow sudden shifts in cargo or steering, leading to an accident.

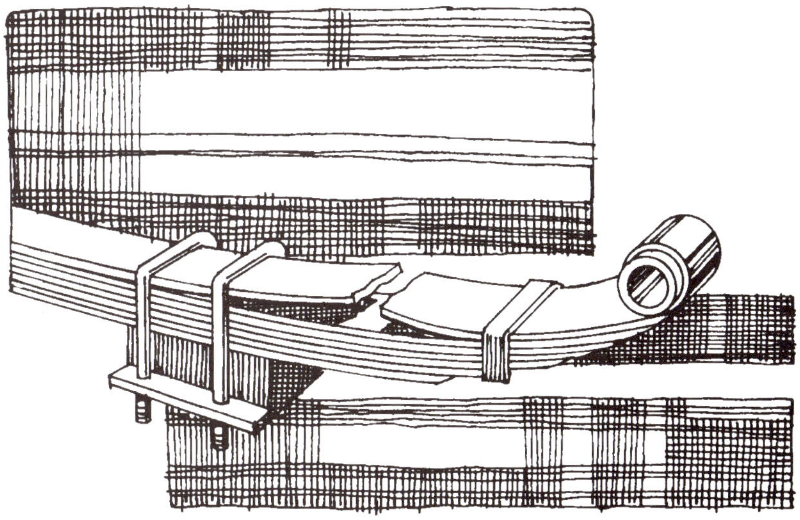

You should check for loose spring hangers that allow movement of an axle from its proper position. Look to see that spring hangers aren't cracked or broken. Torque rotor arm, U-bolts, and other axle positioning parts must be checked for cracks or damage.

Check for missing or broken leaves in any leaf spring. Regulations say if a fourth or more of the leaves are missing or broken, it will put the vehicle out of service, but any defect could be dangerous. Broken leaves in a multi-leaf spring or leaves that have shifted might hit a tire or other part. Look too for shock absorber leaks. Air suspension parts should not show signs of damage or leaks. Finally, check for any loose, cracked, broken or missing frame members.

exhaust system

Exhaust system defects can let poison fumes into the cab or sleeper berth. Loose, broken or missing exhaust pipes, mufflers, tailpipes, vertical stacks, mounting brackets, clamps, nuts or bolts all point to potential problems. Make sure exhaust system parts are not rubbing against fuel system parts, tires or other moving parts of the vehicle. Check for leaks in the exhaust system.

coupling system

Coupling system defects can cause serious accidents or cargo damage. You should check for excess slack or damage of any kind in the fifth wheel locking mechanism. The fifth wheel assembly needs to be inspected for cracks or breaks. The slide mechanism of sliding fifth wheels should not be missing pins or have other defects. Kingpins must be checked for cracking, bending and excessive wear. Missing U-bolts, cracked or broken welds, or other defects in fifth wheel mounting devices signal the need for repair or replacement.

safety equipment

Vehicles should be equipped with certain emergency equipment. Check that a proper fire extinguisher is on board. You should have spare electrical fuses, unless your vehicle is equipped with circuit breakers. Warning devices, such as three reflective warning triangles, are required.

You must inspect cargo for overloading, correct balance and securement before each trip. If the cargo contains hazardous materials, you have to inspect for the proper papers and placarding.

A Seven-Step Pretrip Inspection Method

You should do a pretrip inspection the same way each time so you'll learn all the steps and be less likely to forget something. The following seven-step procedure will be a useful guide and reference.

experience speaking

"That is a necessity on the part of the driver, if he wants to do his job safely. Would you get in my personal car and drive across country without checking it? Same thing goes. I wouldn't come up here and get in these people's vehicle and leave their yard without checking it. Maybe the last driver brought it in with a flat tire on the dual and didn't even know it. But he was going home, so he parked it. I get in it and go out on the street, and I've got a flat. So naturally everything would definitely have to be checked."

Talmage
Driver
Tennessee

vehicle overview

Approach the vehicle. Notice its general condition. Look for damage or if the vehicle is leaning to one side. Look under the vehicle for fresh oil, coolant, grease or fuel leaks. Check the area around the vehicle for hazards to vehicle movement (people, other vehicles, objects, low hanging wires or limbs and branches).

Review the last vehicle inspection report. Drivers may have to make a vehicle inspection report in writing each day. The vehicle owner should repair any items in the report that affect safety. You should look at the last report to find out what, if anything, was the matter. Inspect the vehicle to find out if problems were fixed.

engine compartment

Check that the parking brakes are on and wheels chocked. You may have to raise the hood, tilt the cab (secure loose things so they don't fall and break something) or open the engine compartment hood. Check the following:

Engine oil level.

Coolant level in radiator; condition of hoses.

Power steering fluid level; hose condition (if so equipped).

Windshield washer fluid level.

Battery fluid level, connections and tie downs (battery may be located elsewhere).

Automatic transmission fluid level (may require engine to be running).

Check alternator, water pump, air compressor belts, air conditioning for tightness and excessive wear — Be prepared by learning how much give the belts should have when adjusted right, and check each one.

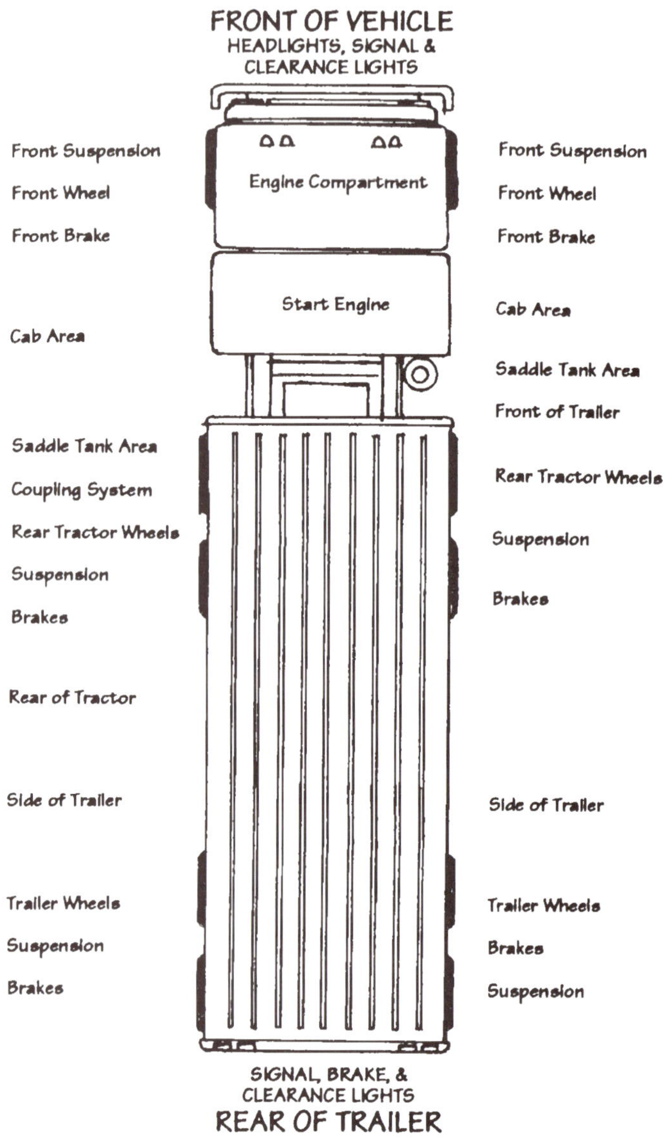

Leaks in the engine compartment (fuel, coolant, oil, power steering fluid, hydraulic fluid, battery fluid).

Cracked or worn electrical wiring insulation.

Check the air brake air compressor drive belt (if compressor is belt driven). If the air compressor is belt-driven, check the condition and tightness of the belt. The belt should be in good condition.

Lower and secure the hood, cab, or engine compartment door.

Check both wet and dry air reservoirs by opening the petcocks and then close petcocks again.

Double check all primary and secondary latches.

inspect inside the cab

Get in and start the engine. Make sure the parking brake is on. Put gearshift in neutral (or park if automatic). Start the engine; listen for unusual noises.

Look at the gauges:

Oil pressure should come up to normal within seconds after the engine is started.

Ammeter and/or voltmeter should be in normal range(s).

Coolant temperature should begin gradual rise to normal operating range.

Engine oil temperature should begin gradual rise to normal operating range.

Air pressure should build to governed cut-out pressure (100-125 psi).

Oil, coolant, charging circuit warning lights should go out right away.

Check condition of controls. Check all of the following for looseness, sticking, damage or improper setting.

Steering wheel.

Clutch.

Accelerator.

Brake controls:

Foot brake;

Trailer brake;

Parking brake;

Front brake limiting valve if equipped;

Tractor protection valve;

Retarder controls (if vehicle has one).

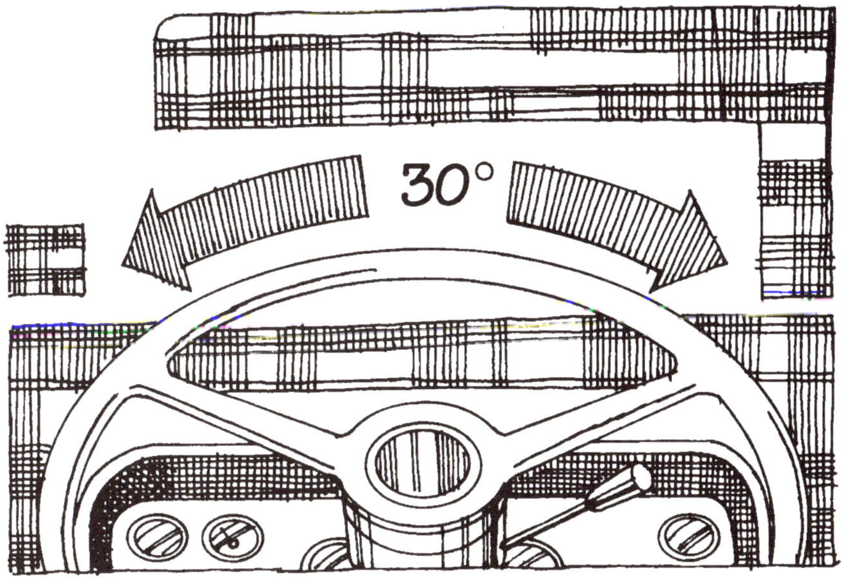

Transmission controls.

Interaxle differential lock (if vehicle has one).

Horn(s).

Windshield wiper / washer.

Lights:

Headlights;

Dimmer switch;

Turn signal;

Four-way flashers;

Clearance, identification, marker light switch(s).

Inspect mirrors and windshield for cracks, dirt, illegal stickers or other obstructions to seeing. Clean and adjust as necessary.

Check Emergency Equipment.

Check for safety equipment:

Spare electrical fuses (unless vehicle has circuit breakers);

Three red reflective triangles;

Properly charged and rated fire extinguisher.

Optional items such as:

Tire chains (where winter conditions require them);

Tire changing equipment;

List of emergency phone numbers;

Accident reporting kit (packet);

First aid kit.

check lights

Make sure the parking brake is set, turn off the engine, and take the key with you. Turn on headlights (low beams) and four-way flashers, and get out. Go to the front of the vehicle and check that low beams are on and both of the four-way flashers are working. Push dimmer switch and check that high beams work.

walkaround inspection

Turn off the headlights and four-way hazard warning flashers. Turn on the parking, clearance, side-marker and identification lights. Turn on right turn signal, and start walkaround inspection.

General

Walk around and inspect.

Clean all lights, reflectors and glass as you go along.

Left Front Side

Driver's door glass should be clean.

Door latches or locks work properly.

Left front wheel.

Condition of wheel and rim – missing, bent, broken studs, clamps, lugs, and signs of misalignment.

Condition of tires -- properly inflated, valve stem and cap okay, no serious cuts, bulges, tread wear.

Use wrench to test rust streaked lug nuts, indicating looseness.

Hub oil level okay, no leaks.

Left front suspension.

Condition of spring, spring hangers, shackles, U-bolts.

Shock absorber condition.

Left front brake.

Condition of brake drum.

Condition of hoses.

Front

Condition of front axle.

Condition of steering system.

No loose, worn, bent, damaged or missing parts.

Must grab steering mechanism to test for looseness.

Condition of windshield.

Check for damage, and clean if dirty.

Check windshield wiper arms for proper spring tension.

Check wiper blades for damage, stiff rubber and securement.

Lights and reflectors.

Parking, clearance and identification lights clean, operating and proper color (amber at front).

Reflectors clean and proper color (amber at front).

Right front turn signal light clean, operating and proper color (amber or white on signals facing forward).

Right Side

Right front: check all items as done on left front.

Primary and safety cab locks engaged (if cab-over-engine design).

Right fuel tank(s).

Securely mounted, not damaged or leaking.

Fuel crossover line secure.

Tank(s) contain enough fuel.

Cap(s) on and secure.

Condition of visible parts.

Rear of engine – not leaking.

Transmission -- not leaking.

Exhaust system -- secure, not leaking, not touching wires, fuel or air lines.

Frame and cross members – no bends, cracks.

Air lines and electrical wiring -- secured against snagging, rubbing, wearing (not touching catwalk, etc.).

Spare tire carrier or rack not damaged (if so equipped).

Spare tire and/or wheel securely mounted in rack.

experience speaking

"Proper inspections keep you out of trouble with the DOT, probably first and foremost. They're inspecting everywhere. The states are picking up the program. Rather than have trucks tied up, and the cost of being tied up and at somebody else's mercy with repairs, it's better to do it at home. Have it repaired in a shop where it's both close and cheaper to us. It serves to expedite the load, have it delivered on time rather than being tied up. There are many cost savings, including fuel economy and just general maintenance savings to have service done at home."

Carl
Safety Engineer
Wisconsin

Spare tire and wheel (proper size, properly inflated).

Coupling System Areas

Check fifth wheel (lower).

Securely mounted to frame.

No missing, damaged parts.

Enough grease.

No visible space between upper and lower fifth wheel.

Locking jaws around the shank, not head of kingpin.

Release arm properly seated and safety latch/lock engaged.

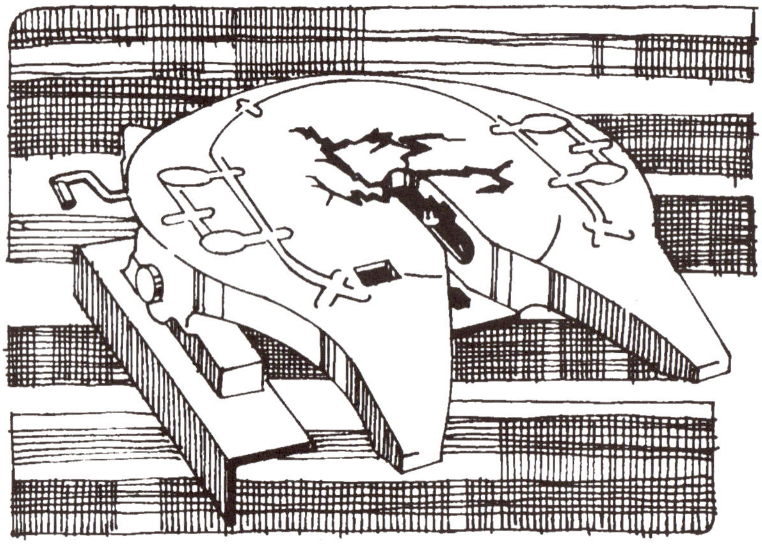

Fifth wheel (upper).

Glide plate securely mounted to trailer frame.

Kingpin not damaged.

Air and electric lines to trailer.

Electrical cord firmly plugged in and secured.

Air lines properly connected to glad hands, no air leaks, properly secured with enough slack for turns.

All lines free from damage.

Sliding fifth wheel.

Slide not damaged or parts missing.

Properly greased.

All locking pins present and locked in place.

If air powered – no air leaks.

Check that fifth wheel is not so far forward that tractor frame will hit landing gear, or cab will hit the trailer, during turns.

Right Rear

Condition of wheels and rims – no missing, bent, broken spacers, studs, clamps, lugs.

Condition of tires – properly inflated, valve stems and caps okay, no serious cuts, bulges, tread wear, tires not rubbing each other and nothing stuck between them.

Tires same type, not mixed radial and bias types.

Tires evenly matched (same sizes).

Wheel bearing / seals not leaking.

Suspension.

Condition of spring(s), spring hangers, shackles and U-bolts.

Axle secure.

Powered axle(s) not leaking lube (gear oil).

Condition of torque rod arms, bushings.

Condition of shock absorber(s).

If retractable axle equipped, check condition of lift mechanism, if air powered check for leaks.

Brakes.

Condition of brake drum(s) and shoes (minimum of 1/4-inch).

Condition of hoses – look for any wear due to rubbing.

Slack adjusters.

Spring brakes.

Lights and reflectors.

Side-marker lights clean, operating and proper color (red at rear, others amber).

Side-marker reflectors clean and proper color (red at rear, others amber).

Rear

Lights and reflectors.

Rear clearance and identification lights clean, operating and proper color (red at rear).

Reflectors clean and proper color (red at rear).

Tail lights clean, operating and proper color (red at rear).

Right rear turn signal operating, and proper color (red, yellow or amber at rear).

Doors and hinges.

License plate(s) present, clean and secured.

Splash guards present, not damaged, properly fastened, not dragging on ground or rubbing tires.

Landing Gear

Fully raised, no missing parts, not bent or otherwise damaged.

Crank handle in place and secured.

If power operated, no air or hydraulic leaks.

Double and Triple Trailers

Shut-off valves (at rear of trailers, in service and emergency lines):

Front trailers: OPEN.

Last trailer: CLOSED.

Converter dolly air tank drain valve: CLOSED.

Be sure air lines are supported and glad hands are properly connected.

If spare tire is carried on converter gear (dolly), make sure it's secured.

Be sure pintle-eye of dolly is in place in pintle hook of trailer(s).

Make sure pintle hook is latched.

Safety chains should be secured to trailer(s).

Be sure light cords are firmly in sockets on trailers.

Left Side

Check all items as done on right side, plus:

Battery (if not mounted in engine compartment);

Battery box securely mounted to vehicle;

Box has secure cover;

Battery(s) secured against movement;

Battery(s) not broken or leaking;

Fluid in battery(s) at proper level (except maintenance-free type);

Cell caps present and securely tightened (except maintenance-free type);

Vents in cell caps free of foreign material (except maintenance-free type).

check signal lights

Get in and turn off lights.

Turn off all lights.

Turn on stop lights (apply trailer hand brake, or have a helper put on the brake pedal).

Turn on left turn signal lights.

Get Out and Check Lights.

Left front turn signal light clean, operating and proper color (amber or white on signals facing the front).

Left rear turn signal light and both stop lights clean, operating and proper color (red, yellow, or amber).

final brake system check

Test Parking Brake

Stop the vehicle, put the tractor parking brake on (yellow four-sided button), and gently pull against it in a low gear to test that the parking brake will hold.

Test Trailer Emergency Brakes

Charge the trailer air brake system and check that the trailer rolls freely. Then stop and pull out the tractor protection valve control, or place it in the emergency position (red eight-sided button). Pull gently on the trailer with the tractor to check that the trailer emergency brakes are on.

Test Service Brakes

Wait for normal air pressure, release the parking brake, move the vehicle forward slowly (about five mph), and apply the brakes firmly using the brake pedal. Note any vehicle pulling to one side, unusual feel or delayed stopping action.

Test Trailer Service Brakes

Check for normal air pressure, release the parking brakes, move the vehicle forward slowly and apply trailer brakes with the hand control (trolley valve), if so equipped. You should feel the brakes come on. This tells you the trailer brakes are connected and working. (The trailer brakes should be tested with the hand valve but controlled in normal operation with the foot pedal, which applies air to the service brakes at all wheels.)

Test Air Leakage Rate

Chock truck first. With a fully-charged air system (typically 125 psi), turn off the engine, release the service brake, and time the air pressure drop. The loss rate should be less than two psi in one minute for single vehicles, less than three psi in one minute for combination vehicles. Then apply 90 psi or more with the brake pedal. After the initial pressure drop, if the air pressure falls more than three psi in one minute for single vehicles (more than four psi for combination vehicles), the air loss rate is too much. Check for air leaks and fix them before driving the vehicle. Otherwise, you could lose your brakes while driving.

Test Low Pressure Warning Signal

Shut the engine off when you have enough air pressure (low pressure warning signal not on). Turn the electrical power on, and step on and off the brake pedal to reduce air tank pressure. The low air pressure warning signal must come on before the pressure drops to less than 60 psi in the air tank (or tank with the lowest air pressure, in dual air systems).

If the warning signal doesn't work, you could lose air pressure, and you wouldn't know it. This could cause sudden emergency braking in a single circuit air system. In dual systems, the stopping distance will be increased. Only limited braking can be done before the spring brakes come on.

Check that the Spring Brakes Come on Automatically

Chock the wheels, release the parking brakes when you have enough air pressure to do it, and shut the engine off. Step on and off the brake pedal to reduce the air tank pressure. The parking brake knob should pop out when the air pressure falls to the manufacturer's specification (usually in a range between 20-40 psi). This causes the spring brakes to come on.

Test Tractor Protection Valve

Charge the trailer air brake system. (That is, build up normal air pressure and push the air supply knob in). Shut the engine off. Step on and off the brake pedal several times to reduce air pressure in the tanks. The tractor protection valve control (also called the trailer air supply control) should pop out (or go from normal to emergency position) when air pressure falls into the pressure range specified by the manufacturer. (Usually within the range of 20 to 45 psi). If the tractor protection valve doesn't work right, an air hose or trailer brake leak could drain all the air from the tractor. This would cause the emergency brakes to come on, with possible loss of control.

Check Rate of Air Pressure Buildup

With the engine at operating RPM, the pressure should build up from 85 to 100 psi within 45 seconds in dual air systems. (If the vehicle has larger than minimum air tanks, the buildup time can be longer and still be safe. Check the manufacturer's specifications.) In single air systems (pre 1975), typical requirements are pressure buildup from 50 to 90 psi within three minutes with the engine at an idle speed of 600-900 RPM. If air pressure doesn't build up fast enough, your pressure may drop too low during driving, requiring an emergency stop. Don't drive until you get the problem fixed.

Check Air Compressor Governor Cut-in and Cut-out Pressures

Pumping by the air compressor should start at about 90-100 psi and stop at about 125 psi. (Check manufacturer's specifications.) Run the engine at a fast idle. The air governor should cut-out the air compressor at about the manufacturer's specified pressure. The air pressure shown by your gauge(s) will stop rising. With the engine idling, step on and off the brake to reduce the air tank pressure. The compressor should cut-in at about the manufacturer's specified cut-in pressure. The pressure should begin to rise. If the air governor doesn't work as described above, it may need to be fixed. A governor that doesn't work right may not maintain enough air pressure for safe driving.

Double and Triple Trailers (Check that Air Flows to All Trailers)

Use the tractor parking brake and chock the wheels to hold the vehicle. Wait for air pressure to reach normal, then push in the red trailer air supply knob. This will supply air to the emergency (supply) lines. Use the

experience speaking

"We would expect that our drivers would inspect the vehicle each time they have a significant stop. It's a pretty common thing that load securing devices do not stay necessarily as you put them when you loaded the thing. From that standpoint, each time our driver makes a meal stop, an overnight stop, particularly, he checks his chains to make sure that they're still tight, that the vehicles are still secured in the manner that they were put on the load originally. It's absolutely essential that the unit be completely inspected after an overnight stop, for example. We found instances where we've had some electrical problems that crept in overnight, even moisture, whatever."

Gilbert
Safety Director
Georgia

trailer hand brake to provide air to the service line. Go to the rear of the rig. Open the emergency line shut-off valve at the rear of the last trailer. You should hear air escaping, showing the entire system is charged. Close the emergency line valve. Open the service line valve to check that service pressure goes through all the trailers (this test assumes that the trailer hand brake or the service pedal is on), then close the valve. If you don't hear air escaping from both lines, check that the shut-off valves on the other trailer(s) and dolly(s) are in the OPEN position. You must have air all the way to the back for all the brakes to work.

This completes the pretrip inspection. If you find anything unsafe during the pretrip inspection, **GET IT FIXED**. State and Federal laws forbid operating an unsafe vehicle. And your good judgment should echo that sentiment.

On-The-Road Inspections

check vehicle operation regularly

You should check:

- Instruments;
- Air pressure gauge;
- Temperature gauges;
- Pressure gauges;
- Ammeter/voltmeter;
- Mirrors;
- Tires;
- Cargo, cargo covers and cargo restraints.

If you see, hear, smell, or feel anything that might mean trouble, check it out. Be aware of the applicable federal regulations.

safety inspection

Drivers of trucks and truck tractors must inspect within the first 25 miles of a trip and every 150 miles or every three hours (whichever comes first) afterward.

Check the following things.

Cargo doors and/or cargo securement.

Tires – enough air pressure; not overheated.

Brakes – not overheated (put back of hand near brake drums to test).

Coupling devices.

After Trip Inspection

You may have to make a written report each day on the condition of the vehicle(s) you drove. Report anything affecting safety or that might possibly lead to mechanical breakdown.

Drain moisture from air tanks and fill fuel tanks (as required or allowed by your employer).

Identify any problems found during your enroute (on-the-road) inspection or discovered while driving; such as unusual noises or vibrations. Conduct an inspection of your vehicle to further identify or locate these problems and discover any newly developing malfunctions.

Identify and diagnose the source of problems or malfunctions.

Complete an accurate inspection report. The vehicle inspection report tells the vehicle owner about problems that may need fixing. A copy of the driver's vehicle inspection report with the mechanic's signature must be placed on the vehicle for the next driver.

Additional Inspection Procedures for Special Rigs

There are many different types of special vehicles or rigs. Each may require additional, sometimes unique, inspection procedures. While there isn't space in this manual to deal with all the different kinds of special rigs, three common types will be covered here in brief: Twin (multiple) trailers; tankers; and refrigerated units (reefers).

inspecting twin trailers (multiple trailers)

Make normal pretrip inspection.

Give special attention to the following items.

After hook-up, check air lines and valves between the two trailers.

Valve position must be as follows:

Rear of lead trailer - OPEN;

Rear of second trailer - CLOSED;

Converter gear air tank valve - CLOSED.

Be sure air lines are properly supported and glad hands properly connected.

Hook-up should be as follows:

Emergency line at front of converter gear to emergency glad hand at rear of lead trailer;

Emergency line at rear of converter gear to emergency glad hand at front of second trailer;

Make same check of service lines.

Check brake operation making sure that all brakes apply and release properly.

If spare tire is carried on converter gear, make sure it's properly secured.

Be sure pintle-eye of dolly is properly in place in pintle hook of lead trailer.

Make sure pintle hook is properly latched.

Safety chains should be properly secured to points of attachment on lead trailer.

Be sure light cords are firmly in sockets on both front and rear trailers.

Two chock-blocks should be available on unit at all times.

Reinspect unit at each stop (as you would a single trailer) with special attention to the following:

Brake lines and glad hands;

Electrical lines and connections;

Pintle hook latch and safety chains.

inspecting tankers

Make normal pretrip inspection.

Pay special attention to:

Dome covers to tank compartments – closed and secured;

Hoses and pumps – fastened and secured;

Pumps free to turn;

All valves in off position and properly locked;

Condition of tank -- no dents or leaks;

Proper hazardous materials placards displayed (if applicable).

Inspection procedures will depend upon the tanker type and the cargo being transported.

Cargo may vary from asphalt and roofing compound to milk and orange juice, with solvents, glues, acids, wines and chocolate in between.

Because there are so many types of tankers and cargo is so diversified, inspection procedures will vary.

inspecting refrigerated units (reefers)

Make normal pretrip inspection.

Check condition of trailer.

Holes in walls, ceiling or floor.

Damaged cold air circulation ducts.

Doors (properly locked and sealed).

Any additional containers or compartments inside trailer.

Any rails or racks on ceiling (properly secured).

Check condition of refrigerator unit.

Water, oil and coolant level.

Suction pressure in compound gauge.

Check fuel level in reefer fuel tank.

Be sure fuel line from tank to refrigerator unit is properly secured and not leaking.

Thermostat should be set at appropriate temperature (depending on type of cargo being transported).

Test run refrigeration unit for 30 minutes to an hour before loading (or as directed by your employer).

vehicle inspections review

1. What is the most important reason for doing vehicle inspections?
2. Name some key steering system parts.
3. Name some suspension system defects.
4. What is the minimum tread depth for front tires? Of other tires?
5. Name some things you should check on the front of your vehicle during the walkaround inspection.
6. What should wheel bearings be checked for?
7. How many red reflective triangles should you carry?
8. How do you test hydraulic brakes for leaks?
9. What is a dual air brake system?
10. How can you check slack adjusters?
11. How can you test the low pressure warning signal?
12. How can you check that the spring brakes come on automatically?

Basic Control: The Essentials 5

Basic Control 5 – 1

starting, warming up & shutting down the engine

putting the vehicle in motion & stopping

backing in a straight line

turning the vehicle

Basic Control 5 – 1

Before actually going out onto the highway behind the wheel of a large commercial vehicle, you must understand the most basic elements of vehicle control.

Operating a tractor-trailer combination is a far cry from a Sunday drive in the family car or a trip to the lumber yard in a long bed pickup truck. The differences are important and many.

Starting, Warming Up & Shutting Down the Engine

Differences between big rigs and everyday family transportation begin with starting the vehicle. Procedures vary by engine type. Gasoline engines differ from diesel engines. And two-cycle diesel engines differ somewhat from four-cycle types. Know the kind of engine in the vehicle you'll be driving and the basic procedures to be followed for that engine.

starting the two-cycle diesel engine

Apply the parking brake. Place the Stop and Emergency Stop (if so equipped) controls into the Run position.

Depress the clutch, and move the gearshift lever to neutral. Turn the switch-key on. Depress the clutch pedal. This prevents the starting motor from turning the transmission gears.

Operate the starter. If the engine doesn't start in 15 to 20 seconds, turn the starter off for at least one minute, to allow it to cool, then try again.

Control engine speed with the foot throttle until it's running smoothly. Do not "race" the motor. When the engine oil pressure reaches normal operating range, set the hand throttle at 1/3 the normal RPM operating range.

Gradually reduce the hand throttle to idle speed as the engine warms. Check all instruments for system malfunctions.

starting the four-cycle diesel engine

Apply the parking brake. Close the throttle and depress the clutch pedal.

Turn the switch key on. Operate the starter. If the engine doesn't start in 15 or 20 seconds, turn it off. Wait at least one minute and try again.

starting the gasoline engine

Apply the parking brake. Depress the clutch, and move the gearshift lever to neutral. Set the choke. In cold weather, pull the choke all the way out. In warm weather, little or no choke will be required.

Turn the ignition switch on. Operate the starter (no more than 15 to 30 seconds at a time). Check all instruments once the engine has started.

Control the engine with the foot throttle and choke until it begins to run smoothly. Set the hand throttle at fast idle.

For fuel efficient starting, don't pump the throttle. This squirts gas into the cylinder, which is not burned when the engine fires. Don't "rev" the engine. Don't leave the choke on any longer than necessary.

engine warmup

Engine warmup is generally done at a low RPM level (usually between 800 and 1,000 RPM). The purpose of warming up the engine to the low side of normal operating temperature is to allow oil to warm and circulate, and oil pressure to build up.

During warmup, the cylinder walls become coated. Bearings are lubricated. Coolant temperature increases. Air pressure builds up.

Check the vehicle's operator's manual for recommended idling time and whether warmup while moving is recommended or not. Usually a few minutes of idling time is the maximum required and recommended for sufficient warmup.

When driving during warmup, keep RPM low. Avoid high RPM when accelerating in first gear. Shift to the next highest gear as soon as possible. These are the keys to a fuel efficient warmup.

There are consequences for not warming up the engine.

Rapid acceleration or "over revving" prior to warmup can cause damage to moving parts that haven't been sufficiently lubricated. There may be crankshaft and bearing damage.

When all instrument readings reach the normal range, operate at cruising speeds for whatever gear you're in. Keep acceleration smooth and easy.

engine shutdown

A cooldown period can be as important as a proper warmup. During cooldown, the engine should idle at least five minutes. Coolant and oil flow at reduced temperatures as the engine idles. Heat generated during normal operation is dissipated.

The ideal cooldown period will vary according to manufacturer's specification, the nature of the trip completed and the nature of the load hauled. Engine damage could occur without sufficient cooldown.

The cooldown period isn't as critical an element in shutting down a gas engine.

Steps in engine shutdown vary slightly. The information that follows covers general good practice. The cooldown period is the key step.

The basic steps in turning a diesel engine off begin with depressing the clutch and moving the gearshift to neutral. Cool the engine down briefly. Turn the engine off. If the engine has a stop control, move it to off. This cuts off fuel flow from injectors.

excessive idling

A visit to any truckstop or other area where vehicles are parked for extended periods might lead you to believe that long periods of idling are not harmful. This is not the case.

Avoid excessive idling. As a rule, today's engines don't need to idle beyond five minutes. Many drivers waste fuel with excessive idling, but carriers often track excessive idling via tachograph and other measures.

Putting the Vehicle In Motion & Stopping

Putting a tractor-trailer combination in motion requires practice because it differs from any other kind of vehicle.

testing tractor-trailer hookup

Coupling must be tested every time a trailer is hooked up.

- Vehicles equipped with independent trailer brake control.
  - Depress clutch, move gearshift to lowest forward gear.
  - Engage independent trailer brake control, applying trailer brakes.
  - Release clutch to friction point.
  - Pull forward gently against locked trailer brakes.
  - Disengage clutch.
  - Repeat procedure.

  Note: This will also show whether trailer brakes are working or not.
- Vehicles without independent trailer brake control (hand valve).
  - Depress clutch, move gearshift to lowest forward gear.
  - Switch tractor protection valve from Normal to Emergency.
  - Partially engage clutch, and pull forward gently against locked trailer brakes.
  - Repeat procedure.

putting a tractor-trailer in motion

- Depress clutch fully, place transmission in lowest forward gear.
- Increase engine RPM slightly (gas engines only).
- Release clutch to friction point.
  - Point where clutch engages.

Decrease in RPM is a sign that friction point has been reached.

Gradually release clutch until it's fully engaged.

When vehicle starts to move (diesel engines only) gradually increase engine RPM to increase vehicle speed.

When vehicle is in motion with clutch fully engaged, take left foot off clutch until ready to shift again or stop.

Avoid excessive clutch slipping: Engage as quickly as possible without stalling engine.

stopping a tractor-trailer

Release accelerator pedal.

Depress brake pedal for optimum valve opening for stopping rate desired.

As the vehicle begins to slow, decrease brake pedal pressure.

Stop vehicle smoothly by releasing brake pressure in relation to vehicle stopping rate.

As vehicle halts, release pedal completely.

Depress clutch pedal as vehicle halts.

After halting, reapply brake just enough to keep vehicle stationary. Release clutch pedal. Put gearshift lever in neutral if you're parking the truck.

Backing In A Straight Line

Backing is a difficult maneuver to master. Detailed backing maneuvers are covered in greater detail later under the heading "Backing" (5-3). The information presented here is intended only to acquaint you with the procedure of straight line backing. Obviously, straight line backing is the most simple of all backing maneuvers (that is, if any backing maneuver can be called simple).

Step 1: Position vehicle properly.

Put vehicle in position by moving forward until the tractor-trailer is aligned and front (steering) wheels are straight.

experience speaking

"I suppose the worst thing in town is right hand corners. People crowd in on the right side, because you've got to move left pulling a long box like this. I suppose that's the worst thing I'll ever do in traffic is right hand turns. Backing up is suicide. You never back up in traffic. I'd say 95% of in-town accidents are right hand turns or some guy going a little too fast for the traffic conditions."

Don
Recruiter
California

Step 2: Speed.

Put the vehicle in reverse.

Back as slowly as possible (use idle speed).

Don't ride clutch or brake pedals.

Step 3: Checking behind.

Constantly check behind when backing.

Keep doors closed, and always check both mirrors.

Make maximum use of mirrors to detect persons or objects crossing into intended path. If unsure, always stop.

Step 4: Steering.

Keep the vehicle on path.

Do not oversteer.

Correcting drifts.

- Turn steering wheel in direction of the drift as soon as it occurs.
- Do not oversteer, only slight movement is required if drift is caught right away.
- Little drifts need little correction; big drifts need big ones.
- Use push-pull method of steering to keep trailer in line.

A simple method of steering to keep the combination in line.

- When trailer gets bigger in mirror, push steering wheel towards that mirror; for example, towards left, if left mirror.
- Immediately straighten steering wheel out again.
- If late in catching larger image of trailer, push wheel more to correct.

The biggest error in push-pull method is not returning wheel to straight ahead position immediately.

Step 5: Pull up and start again, if too far out of position.

Easier to make corrections moving forward.

Turning The Vehicle

There are four basic rules for turning: Know the vehicle; plan in advance; allow for off-tracking; and watch your mirrors.

In addition to the simplified information that follows, more specific detail can be found under the heading "Managing Space" (6-4).

off tracking

Off-tracking is a condition that occurs in all four-wheel and larger vehicles. Essentially, off-tracking describes the process of the rear wheels of a vehicle following a shorter path than the front wheels. The greater the distance between front and rear wheels, the sharper the turn, the more off-tracking increases.

Two key factors determine the off-track of the trailer: The distance between the kingpin and the rear trailer wheels; and the amount of sideways drag of the rear tires.

The effects of off-tracking of the trailer must be taken into consideration when making turns or taking curves.

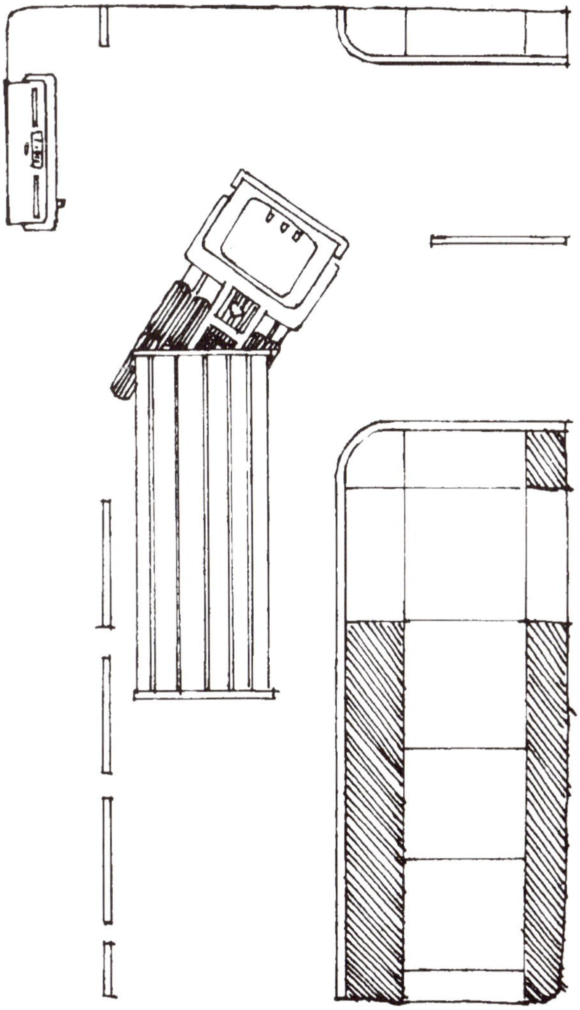

making right turns

As you approach the intersection, adjust your speed. Avoid the need to brake while turning, but do allow for slight acceleration. By shifting into the proper gear before the turn and completing the turn in that same gear, you'll be able to keep both hands on the wheel during the turn. Maintain a light "pull" on the trailer. Never let the trailer push the tractor.

A semi-trailer must pull further into the intersection to avoid crossing over the curb. As a rule of thumb, get approximately 1/2 the length of the rig past the corner point of the intersection before beginning the turn.

When you reach the appropriate point, turn the steering wheel to the right, and complete the turn.

Accelerate slightly to smooth out the turn. During the turn, watch the right mirror for the position of your trailer wheels, as well as vehicles that may have sneaked in beside you and are attempting to pass. Immediately after the turn is complete, turn the steering wheel back to the left to straighten the wheels.

most common right turn errors

Approaching the intersection too fast.

Failure to gear down before a turn.

Shifting gears while turning.

Failure to allow for off-tracking of the rear trailer wheels.

Failure to watch your right mirror before, during and as you complete the turn.

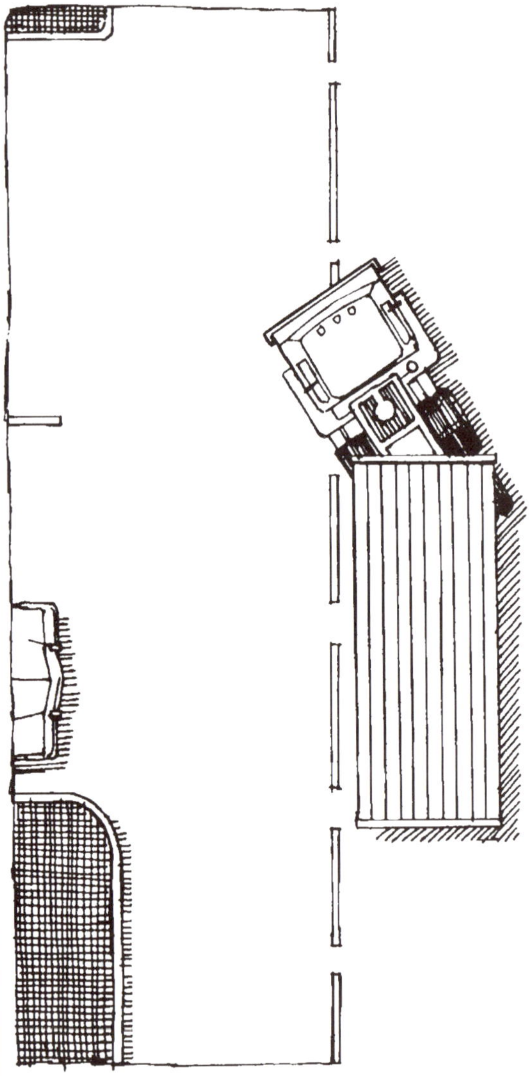

making left turns

Reduce speed as you approach the intersection. Select the proper gear for completing the entire turn. Turn to the left from as far right as possible to allow the trailer plenty of room. If you turn too soon or attempt to turn too tight, the left side of your vehicle may hit another vehicle, because of off-tracking. Watch the left mirror closely before, during and as you complete the turn.

Turn the steering wheel back to the right to straighten the wheels immediately upon completion of the turn.

The most common left turn errors are the same as those for right turns, simply from the other side. If you are on a road with two lanes for turning left, use the outside lane.

curves

Positioning for a curve is important. To set your vehicle up for a right travelling curve, keep the vehicle toward the center of the road **without crossing the center line**. Failure to keep toward the center could result in the rear of your vehicle running off the road.

Approaching a left travelling curve, keep the front of your vehicle as close to the outer edge of the road as possible. This will keep your trailer from running over the centerline on a curve.

basic control review

1. What is the procedure for starting a diesel engine?

2. Why is it important to warm up an engine before heading out?

3. What are some of the dangers of excessive idling?

4. How does turning a tractor-trailer combination vary from turning a straight truck?

5. If stopped on a hill, how can you start moving without rolling back?

6. How often must a vehicle's coupling be tested on a tractor-trailer combination?

7. Straight line backing is the most simple of all backing maneuvers: True or False?

Shifting Gears 5 – 2

shifting controls

shifting gears

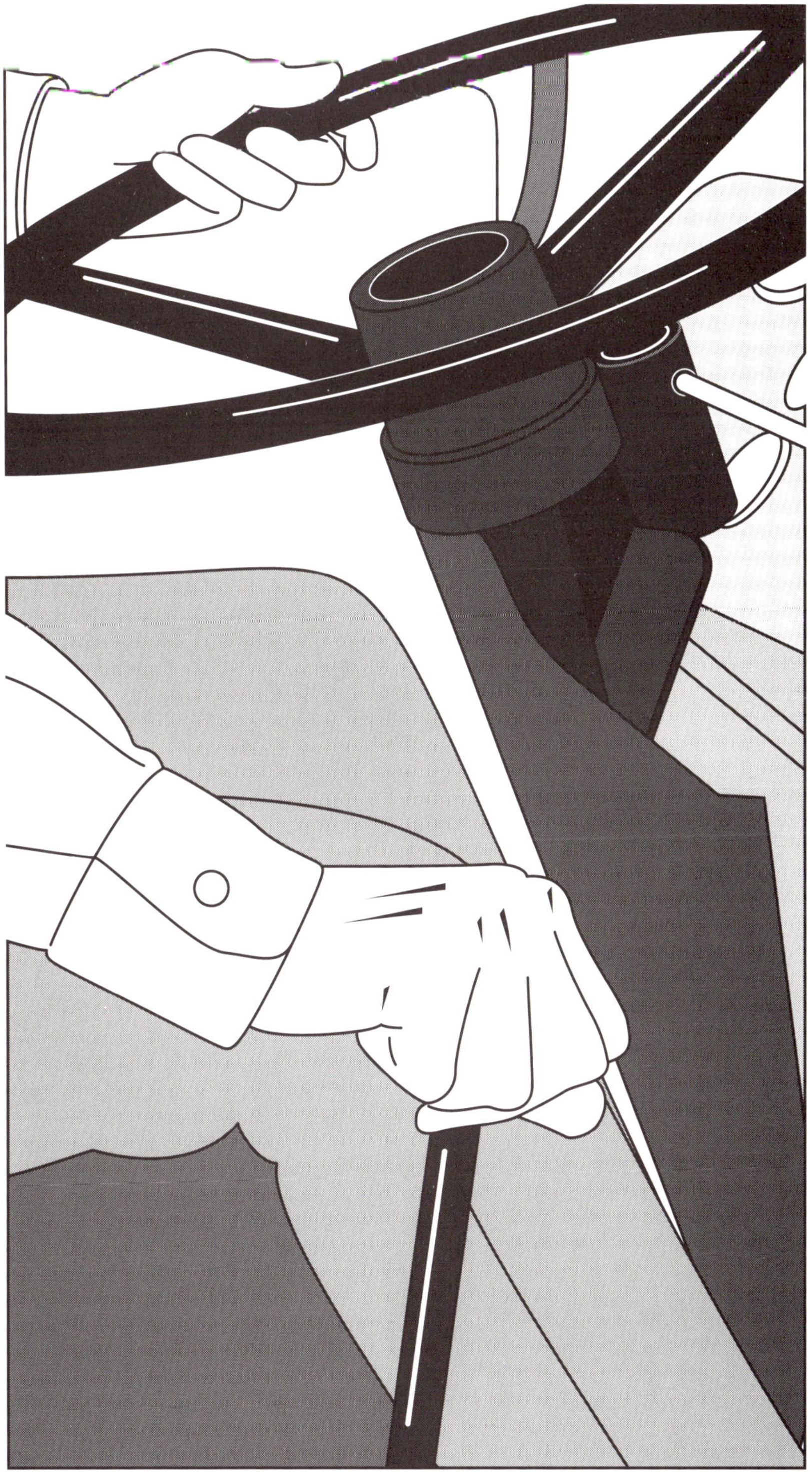

Shifting Gears 5 – 2

Correct shifting of gears is important and often crucial.

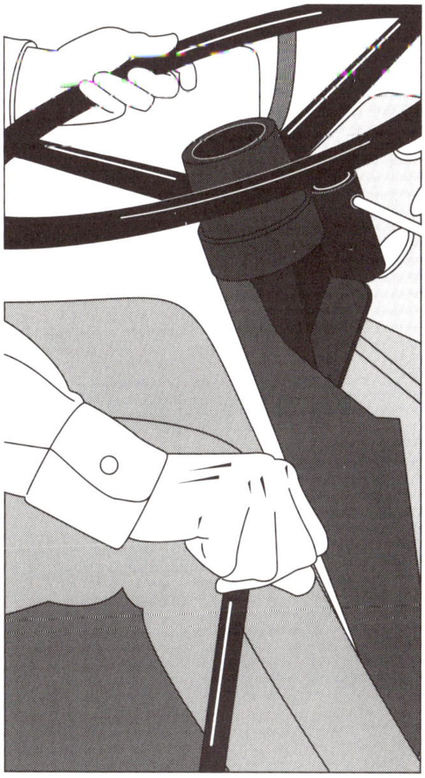

If you can't get your vehicle into the right gear while driving, you will have less control. On a curve or a steep grade, this can mean disaster. Proper shifting involves knowledge and skill. It also involves an understanding of shift controls, when to shift and how to shift.

Shifting Controls

There are three shifting controls: The accelerator; the shift lever and the clutch. The essentials of shifting properly aren't difficult to learn, but it takes much practice and experience to shift correctly in all situations.

accelerator

The accelerator controls the flow of fuel to the engine. That, in turn, determines the speed of the engine in any one gear.

shift lever

The shift lever controls what gear the transmission is in. The gear selected at any given point determines the amount of road speed and power supplied by the engine with the accelerator in any one position. In low gear, there's a lot of power and little road speed. Low gear basically supplies the power needed to build up momentum.

In high gear there's less power supplied, but a lot of road speed. High gear allows the vehicle to obtain and maintain higher road speeds once momentum is developed.

clutch

The clutch is the controlling connection between the engine and the transmission. It determines largely whether the transmission can be shifted safely. With the clutch pedal down, engine and transmission are disengaged, and gears may be

shifted safely. When the clutch pedal is up all the way, engine and transmission are engaged, and gears generally can't be shifted safely.

You can think of the clutch as having three "parts" when driving. The first part is the freeplay, roughly the uppermost 1-2 inches of the stroke. The second is the "working part" of the clutch where it engages the transmission, roughly 5-6 inches of the stroke. The last part, until the pedal hits the floor, is the clutch brake. The clutch brake stops the transmission's rotation. For this reason, you should engage the clutch totally only when the truck is at a dead stop.

coordination

Operation of shifting controls takes coordination and precise timing. Lack of coordination can cause over revving and gear grinding.

There are many different types and brands of transmissions. Each is unique in its own way. It's your responsibility, with the help of the manufacturer's operating instructions and a good teacher, to learn how to shift gears in the vehicles you'll be driving. The information that follows outlines general good practice while driving.

Shifting Gears

manual transmissions: upshifting

Most heavy vehicles with manual transmissions require double clutching to change gears.

This is the basic method:

1. Release accelerator, push in clutch and shift to neutral at the same time;
2. Release clutch;
3. Let engine and gears slow down to the rpm required for the next gear (this takes practice);
4. Push in clutch and shift to the higher gear at the same time;
5. Release clutch and press accelerator at the same time.

Shifting gears using double clutching requires practice. If you remain too long in neutral, you may have difficulty putting the vehicle into the next gear. If so, don't try to force it. Return to neutral, release the clutch, increase engine speed to match road speed, and try again.

Knowing when to shift up. There are two ways of knowing when to shift:

Use engine speed (RPM). Study the driver's manual for your vehicle and learn the operating RPM range. Watch your tachometer, and shift up when your engine reaches the top of the range. (Some newer vehicles use progressive shifting: The RPM at which you shift becomes higher as you move up in the gears.) Find out what's right for the vehicle you'll operate.

Use road speed (m.p.h.). Learn what speeds each gear is good for. Then, by using the speedometer, you'll know when to shift up. With either method, you may learn to use engine sounds to know when to shift.

experience speaking

"A lot of them, I've noticed, speed shift. Their tractors are always jumping. Every gear they go into, it jumps. They aren't shifting properly. There isn't any sense in speed shifting. Vehicles are made for regular shifting and they still do a good job."

Vern
Driver
Wisconsin

manual transmission: downshifting

This is the basic method for downshifting:

1. Release accelerator, push in clutch and shift to neutral at the same time;
2. Release clutch;
3. Press accelerator, increase engine and gear speed to the RPM required in the lower gear;
4. Push in the clutch and shift to lower gear at the same time;
5. Release clutch and press accelerator at the same time.

Downshifting, like upshifting, requires knowing when to shift. Use either the tachometer or the speedometer and downshift at the right RPM or road speed.

Before starting down a hill, slow down, and shift down to a speed that you can control without using the brakes hard. Otherwise the brakes can overheat and lose their braking power. Downshift before starting down the hill. Make sure you're in a low enough gear, usually lower than the gear that was required to climb the same hill.

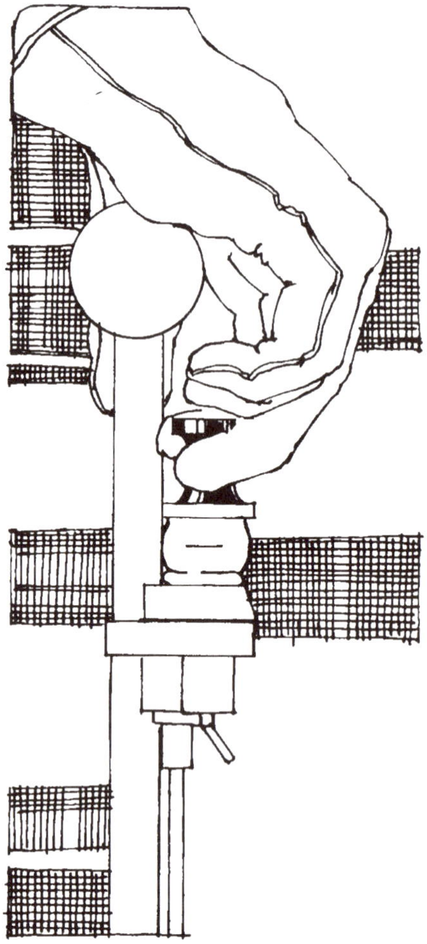

Before entering a curve, slow down to a safe speed, and downshift to the right gear. This lets you use some power through the curve to help the vehicle be more stable while turning. It also lets you speed up as soon as you're out of the curve.

multi-speed rear axles and auxiliary transmissions

Multi-speed rear axles and auxiliary transmissions are used on many vehicles to provide extra gears. You usually control them by a selector knob or switch on the gearshift lever of the main transmission. There are many different shift patterns. Learn the right way to shift gears in the vehicle you'll drive.

automatic transmissions

Some vehicles have automatic transmissions. You can select a low range to get greater engine braking when going down grades. The lower ranges prevent the transmission from shifting up beyond the selected gear (unless the governor RPM is exceeded). It's very important to use this braking effect when going down grades.

retarders

Some vehicles have retarders. Retarders help slow a vehicle, reducing the need for using your brakes. They reduce brake wear and give you another way to slow down. There are many types of retarders (exhaust, engine, hydraulic, electric). Retarders can be controlled by the driver. On some, the retarding power can be adjusted. When turned on, retarders apply their braking power (to the drive wheels only) whenever you let up on the accelerator pedal all the way.

Caution: When your drive wheels have poor traction, the retarder may cause them to skid. Therefore you should turn the retarder off whenever the road is wet, icy or snow covered.

shifting gears review

1. What are two special conditions where you should downshift?

2. When should you downshift automatic transmissions?

3. Retarders keep you from skidding when the road is slippery: True or False?

4. What are two ways to know when to shift?

5. What is progressive shifting?

6. Name the three basic shifting controls?

7. What are two ways of knowing when to shift?

Backing 5 – 3

Backing 5 – 3

Backing is among the most basic of tractor-trailer maneuvers. But it's also a maneuver that's difficult to learn and perform.

Learning to back a tractor-trailer with a level of ease and efficiency requires patience and practice to develop good judgmental skills.

Because you can't see everything behind your vehicle, **backing is always dangerous**. As a driver, you'll want to avoid backing whenever you can. When you park, try to park so you'll be able to pull forward when you leave. When you have to back, there are a few simple safety rules. Check clearances before starting. Be sure your vehicle is secured. Leave the cab. Check above. Check to the sides. Check under the vehicle. Check behind your vehicle. Check in front of your vehicle for adequate swing clearance. Back slowly. Back and turn toward the driver's side (sight side) whenever possible. Use a helper whenever possible. You and your helper must agree on signals in advance to avoid confusion. Your helper must be in a position so that you can see him in the rearview mirror at **all** times. If not, stop!

Use your horn and flashers. Put the warning flashers on before shifting into reverse. If your vehicle isn't equipped with a backup alarm, periodically tap on the electric horn. Keep the windows open and the radio off so that you'll hear any noises.

steering

It takes some getting used to, but the rear tractor axle functions as the trailer steering axle. The direction of the rear tractor wheels determines the direction the trailer will travel. Changing the direction of the tractor steers the trailer. Turn the steering wheel in the opposite direction of your desired trailer path.

The process of backing involves a combination of jacking and chasing. Jacking is so named because turning the tractor causes a controlled jackknife; an angle between tractor and trailer. Once the trailer is going in the correct direction, the tractor is turned just enough to follow, or chase, the trailer.

If the front tractor wheels are turned to the left, the trailer will go right. If the front tractor wheels are turned to the right, the trailer will go left. Steering is as simple and as complex as that.

Should you, while backing, drift from your desired path, turn the steering wheel in the direction of the drift. Only slight movement on your part is required as soon as the drift occurs. Don't oversteer.

backing maneuvers

Straight-line backing is the simplest to learn, yet a fundamental in learning all other backing maneuvers.

Alley dock backing involves backing while turning 90 degrees. It occurs at loading docks when you must back off the street or between two vehicles.

Parallel parking involves backing into a space along a curb or dock. While the principal is the same as parallel parking a car, it's the most difficult maneuver to learn with a tractor-trailer.

Jackknife parking refers to the final parked position of your rig. The tractor and trailer are at a 90 degree angle to shorten the length of the rig. This is required under such situations as when the dock is too close to the street, when parking for side loading or unloading, or when there are obstructions.

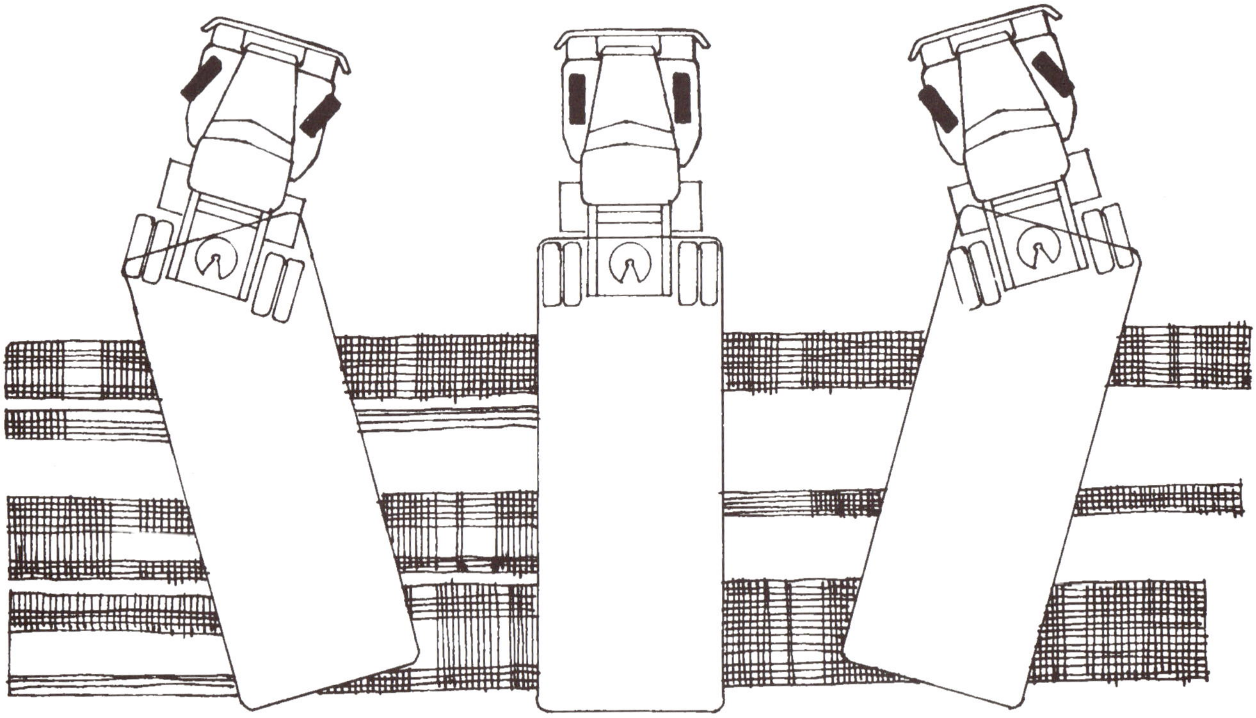

Sight side backing is backing toward the left side of the vehicle. The driver can see along the intended path of the trailer.

Blind side backing is backing toward the right side of the vehicle. You can only see where you're going in your rearview mirrors. Blind side backing is much more difficult than sight side backing, and for safety reasons should always be avoided when possible.

general backing procedures

The key to backing is starting in the proper position. The proper position will vary with the various backing maneuvers. Put the vehicle in position by forward movement. Stop, secure your vehicle, get out and check your position.

Back as slowly as possible using the lowest reverse gear. Don't accelerate. Use idle speed. Don't ride the clutch. Keep off all pedals until stopping.

By looking around and using your mirrors, constantly check behind when backing. Always have your mirrors properly aimed, and use them. Keep your doors closed. (Trailer doors will need to be opened before you're in a position where you don't have room to open them.) Watch behind tractor tires during your jacking sequence. This is particularly important in blind side backing. Often objects, such as parked cars, aren't visible until just before collision. Watch for curbs or other obstacles that might tilt the trailer. Watch for objects overhead, such as wires, pipes, tree limbs and even window air conditioners.

When necessary, start over rather than backing badly. Pull up, and realign the trailer with your desired path whenever your vehicle gets too far out of position. It's always better to pull forward to correct steering errors than to fight it by continuing to back. Make sure you select the path that will align your vehicle properly to its target before pulling forward.

straight-line backing

As mentioned previously, this is the most simple backing maneuver. Position the vehicle, wheels straight and back, making sure not to let the trailer drift to either side. Straight-line backing should be mastered before attempting other backing maneuvers.

experience speaking

"One thing I tell them to do, if you're backing into a long building where the unit will be inside… Normally from a bright sunny day, it's bright outside and you look in there. It's just a black hole. If there are no lights at the dock, take your flashlight, walk back into the building. Lay it up on the dock at the left rear corner where the left rear corner of your trailer will be; where you want the trailer to end up. Then when you get back in your tractor, you've at least got a light. You've got something to guide you back there."

Gene
Instructor
Wisconsin

alley dock backing

Step 1: Approach alley at right angle from right side so traffic can't sneak in.

Check left mirror for people, animals and objects.

Turn to left to pull trailer back end as close to alley as practical.

Check right mirror.

Turn right to pull front of trailer to right, then cut left.

Step 2: With tractor and trailer targeted, turn steering wheel to right when tractor passes center of alley opening.

Step 3: Stop with rig in an arc so you can see left rear corner of trailer and the intended target.

Straighten the steering wheel.

Step 4: Check clearance behind, and to the sides. Open rear doors if there's insufficient clearance to do so later.

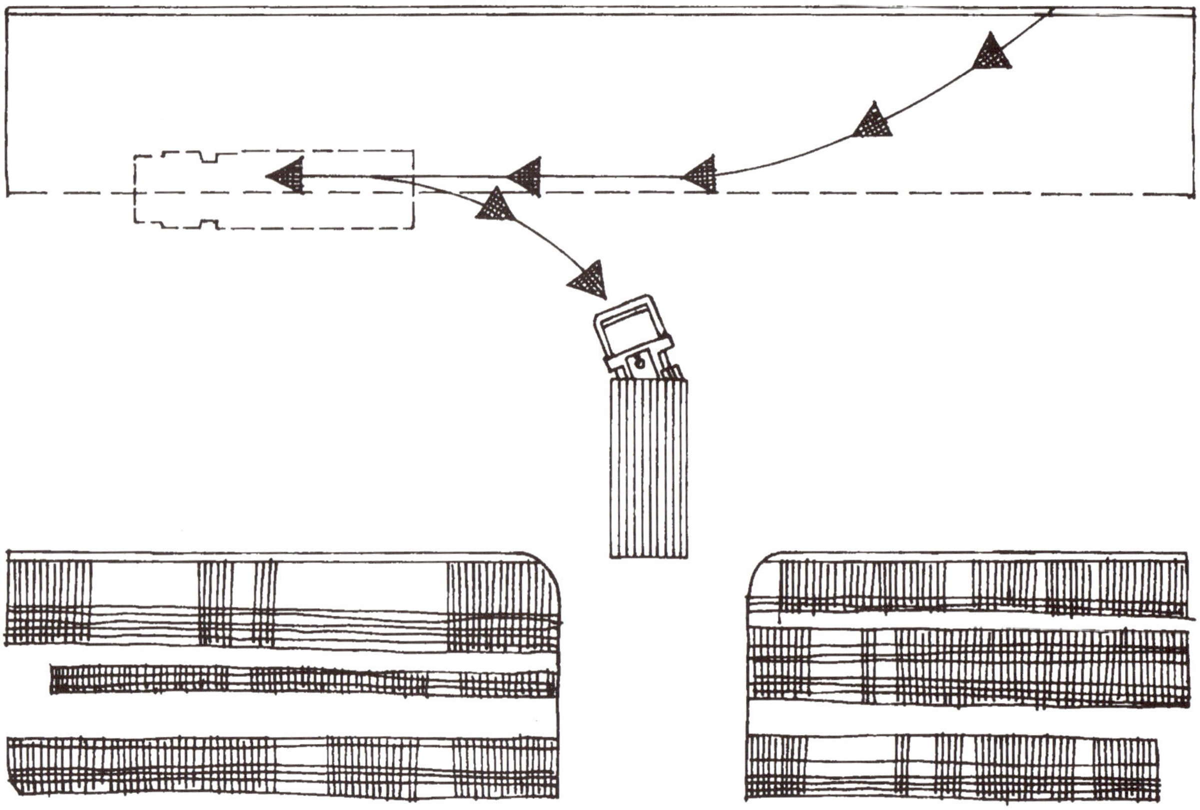

Step 5: Begin backing slowly into alley.

Give warning and continually check path.

Jack tractor right to start trailer along path.

Make adjustments immediately as necessary.

Step 6: Straighten wheels to chase trailer.

As trailer enters alley, put it in position for straight-line backing.

Step 7: Stop, secure vehicle and check behind again.

Walk to dock to check remaining space (if you don't have a qualified helper).

Open rear doors, if you haven't already, before trailer gets to the dock.

Step 8: Ease vehicle into dock.

experience speaking

"If there's any way I can get away from backing on the blind side, I will. Either straight on or backing from the drivers side, or sight side. It's much easier and you can see what's going on. Blind side you just have no concept."

Jim
Driver
Oregon

jackknife parking - sight side

Step 1: Position vehicle properly by pulling past dock while parallel to it.

Step 2: Check clearance.

Step 3: Visualize path of trailer.

Step 4: Follow general backing procedure listed in this section.

Step 5: Watch for rear wheels to reach turning point (approximately midway point of alley). You must anticipate this point or it may be too late.

Step 6: Turn steering wheel to right so trailer is backing in a line that will allow the tractor to follow (approximately at a 15 degree angle).

Step 7: When trailer is about to make contact with dock, turn steering wheel sharply right. (This will put you in the 90 degree jackknife position.)

Caution: **Make sure air and electrical lines are long enough not to be torn off**. Parking at more than a 90 degree angle must be avoided. In addition to the potential for tearing off air and electrical lines, the trailer may move forward and hit the rear corner of the tractor.

jackknife parking - blind side

Never use this maneuver if it can be avoided. If you must blind side back, **always** use a helper.

Step 1: Position vehicle.

Step 2: Check area you'll be backing into. Plan to get out of the truck and check what you are doing every few feet.

Step 3: Visualize path of travel.

Step 4: Follow general backing rules.

Step 5: Watch for rear wheels to reach turning point.

Step 6: Turn steering wheel left, so trailer backs right. Trailer should be backing in a line that allows tractor to follow.

Step 7: When trailer is about to make contact with dock, turn steering wheel sharply to left. (This will put you in the 90 degree jackknife position.)

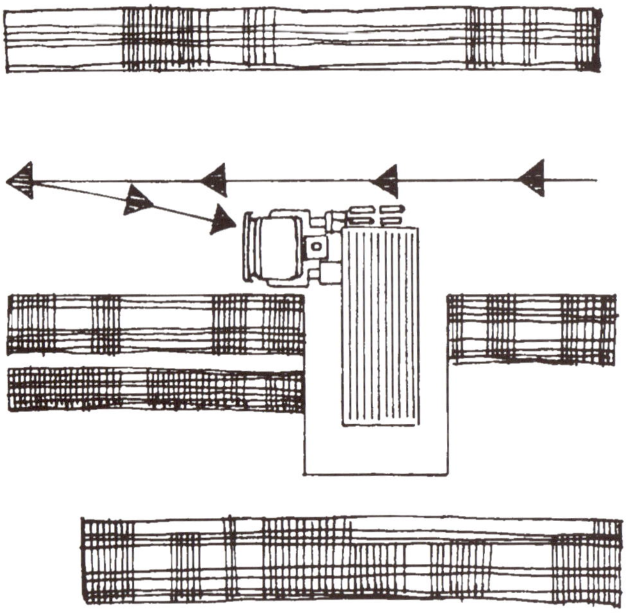

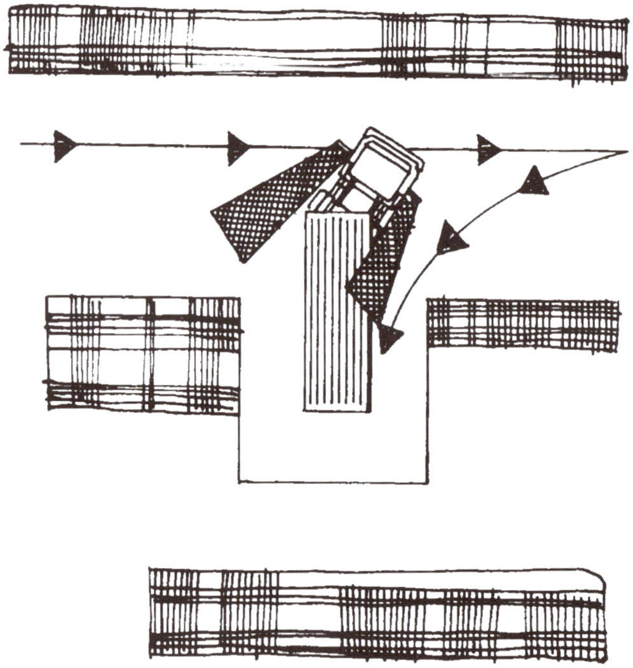

parallel parking

Step 1: Drive past front of parking space, parallel to it. Stop when trailer is approximately three feet from end of space.

Step 2: Check area to be backed into.

Step 3: Start backing, keeping wheels straight.

Step 4: Turn steering wheel slightly to left when vehicle starts backing. Trailer should be entering space at approximately 15 degree angle. Continue backing.

Step 5: Turn steering wheel to left, making sure back right side of trailer is heading toward back right side of space. Continue until front of trailer is even with front of space.

Step 6: Turn steering wheel sharply to right. Continue backing until tractor is even with front of parking space (jackknife position).

Step 7: Turn steering wheel sharply to left. Continue backing until tractor-trailer is within parking space.

Step 8: Pull vehicle forward, adjusting clearance to front and back evenly.

Remember backing is the most difficult procedure to master. Many drivers have spoiled an otherwise excellent safety record with a moment of careless backing. Patience and practice are the key. If you're not sure, don't guess. Stop.

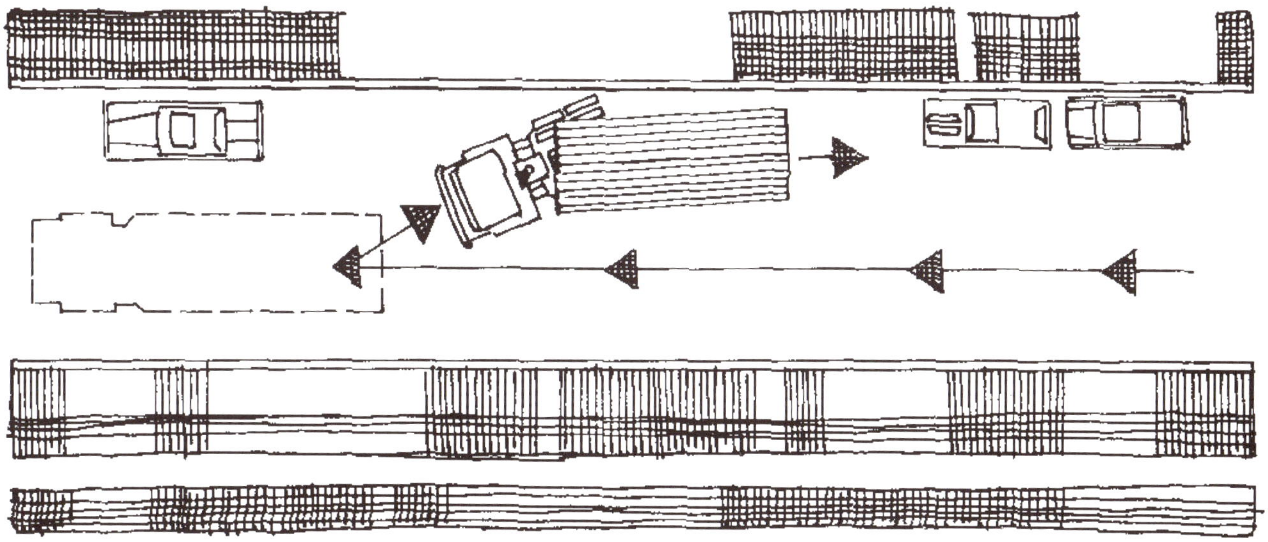

backing review

1. Why should you back toward the driver's side whenever possible?

2. What is a pull up?

3. When backing, why is it important to use a helper?

4. What's the most important hand signal you and your helper should agree on?

5. Backing is seldom dangerous if done properly: True or False?

6. Identify four different backing maneuvers.

Coupling & Uncoupling 5 – 4

coupling tractor-semitrailers

uncoupling tractor-semitrailers

coupling twin trailers

uncoupling twin trailers

other combinations

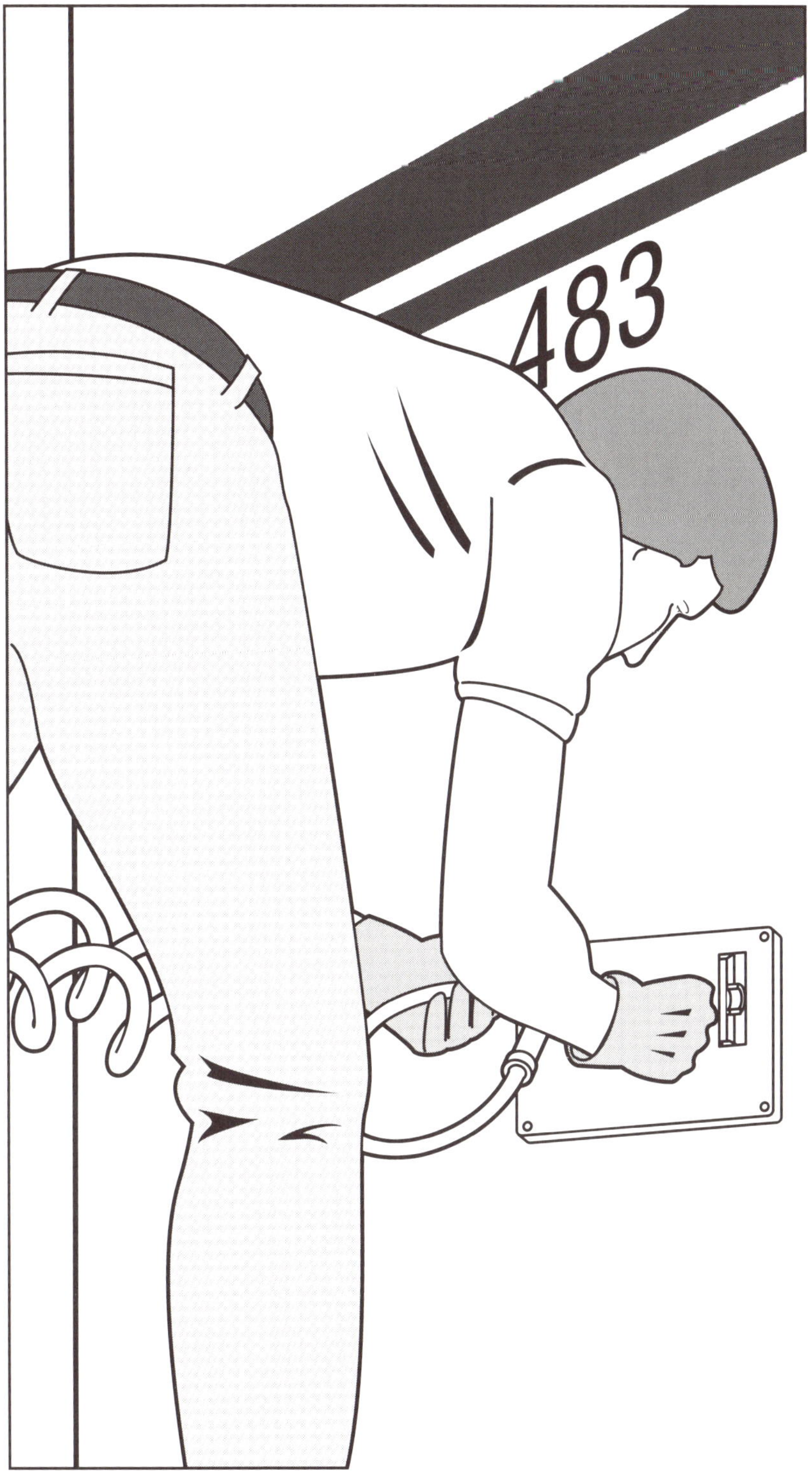

Coupling & Uncoupling 5 – 4

Coupling and uncoupling skills are basic to tractor-trailer operations.

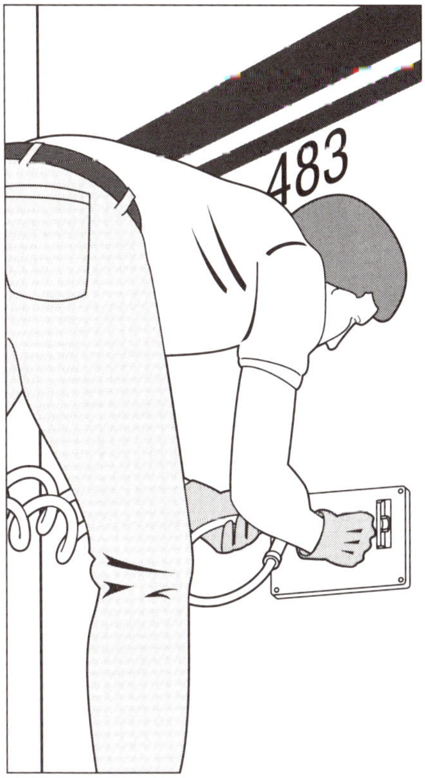

Disaster awaits any driver not well versed in proper coupling and uncoupling procedures.

The FMCSR requires that a driver be skillful in coupling and uncoupling units he'll drive. Every tractor-trailer combination driver must learn how to couple and uncouple by the numbers; that is, each step must be done in exact sequence to avoid injury and damage.

When coupling and uncoupling have been mastered, an experienced driver should be able to couple a tractor-trailer combination in about five minutes if all equipment is operating as it should be and weather conditions are good. When a driver doesn't know what he's doing, serious consequences are likely.

Improper coupling can result in a driver being crushed by rolling wheels or a falling trailer, or even from slamming a tractor against a trailer that's positioned too high or too low for proper hookup. A trailer can be dropped with severe damage to both the trailer and its contents. Air and electrical lines may be damaged or torn off. Landing gear could be damaged or pushed through the trailer floor when a trailer is dropped. The tractor or trailer might roll. Cab damage could be caused by the trailer jumping the pin.

Coupling Tractor-Semitrailers

Step 1: Inspect the fifth wheel.

Check for damaged or missing parts.

Check to see that mounting to tractor is secure; no cracks in frame, etc.

Be sure that the fifth wheel plate is greased as required. Failure to keep the fifth wheel plate lubricated could cause steering problems because of friction between the tractor and trailer.

Check if fifth wheel is in proper position for coupling:

Wheel tilted down towards rear of tractor;

Jaws open;

Safety unlocking handle in the automatic lock position.

If you have a sliding fifth wheel, make sure it's locked.

Make sure the trailer kingpin is not bent or broken.

Step 2: Make sure the area around the vehicle is clear.

Be sure trailer wheels are chocked or spring brakes are on.

Check that cargo (if any) is secured against movement due to the tractor being coupled to the trailer.

Step 3: Position the tractor.

Put the tractor directly in front of the trailer. (Never back under the trailer at an angle. You might push the trailer sideways and break the landing gear or you may push the trailer into something.)

Check position, using outside mirrors, by looking along both sides of the trailer.

Step 4: Back slowly.

Back until the fifth wheel just touches the trailer.

Don't hit the trailer or lift it up onto the fifth wheel.

Step 5: Secure the tractor.

Put on the parking brake.

Put the transmission in neutral.

Step 6: Check trailer height.

The trailer should be low enough that it's raised slightly by the tractor when the tractor is backed under it.

Raise or lower the trailer as needed. (If the trailer is too low, the tractor may strike and damage nose of trailer; if trailer is too high, it may not couple correctly or completely miss the pin.)

Check that the kingpin and fifth wheel are aligned. Make sure jaws are open (if the tractor has been moved any distance.)

Step 7: Connect the air lines to the trailer.

Check glad hand seals, and connect tractor emergency air line to trailer emergency glad hand (usually on the right side when facing the trailer).

Connect service line.

Make sure air lines are safely supported where they won't be crushed or caught while the tractor is backing under the trailer.

Step 8: Supply air to the trailer.

From cab, push in the air supply knob or move the tractor protection valve control from the emergency to the normal position to supply air to the trailer brake system.

Wait until the air pressure is normal.

Check the brake system for crossed air lines.

Shut the engine off so you can hear the brakes.

Apply and release trailer brakes, listen for sound of trailer brakes being applied and released. You should hear the brakes move when applied and air escape when brakes are released.

Check air brake system pressure gauge for signs of major air loss.

When you're sure the trailer brakes are working, start the engine.

Make sure air pressure is up to normal.

Step 9: Lock the trailer brakes in ONE of the TWO following ways:

Move the tractor protection valve control from normal to emergency or pull trailer brake hand control all the way down.

Step 10: Back under the trailer.

Use lowest reverse gear.

Back the tractor slowly under the trailer to avoid hitting the kingpin too hard.

Stop when the kingpin is locked into the fifth wheel.

experience speaking

"I think the biggest thing is hooking up your air lines. Do hook them up and charge them, even if you have spring brakes, it's important to hook the system up, make it functional. And reset them in the event you have broken springs in the spring brakes. That way you won't end up chasing a trailer. The other thing is to make sure that the apron of the trailer hits the fifth wheel and that a percentage or a certain amount of weight is transferred form the dollies onto the tractor so that if the trailer has to move either left or right to slide onto the fifth wheel, not all the stress is handled by the dollies."

John
Driver
New Mexico

Step 11: Check connection for security.

Make sure the trailer landing gear is slightly off the ground.

Pull the tractor gently forward while the trailer brakes are still locked to check that the trailer is locked onto the tractor.

Step 12: Secure vehicle.

Put the transmission in neutral.

Put the parking brakes on.

Shut off engine and take the key with you so someone else won't move the truck while you're under it.

Step 13: Inspect coupling.

Use a flashlight if necessary.

Make sure there's no space between the upper and lower fifth wheel. If there's space, something is wrong (kingpin may be on top of closed fifth wheel jaws; trailer would come loose very easily).

Go under trailer and look into the back of the fifth wheel. Make sure the fifth wheel jaws have closed around the shank of the kingpin.

Check that the locking lever is in the lock position.

Check that the safety catch is in position over the locking lever. (On some fifth wheels, the catch must be put in place by hand.)

If the coupling isn't right, don't drive the coupled unit; get it fixed.

Step 14: Connect the electrical cord and check the air lines.

Plug the electrical cord into the trailer and fasten the safety catch.

Check both air and electrical lines for signs of damage.

Make sure air and electrical lines won't hit any moving parts of the vehicle or rest on the catwalk.

Step 15: Raise trailer supports (landing gear).

Use low gear range (if so equipped) to begin raising the landing gear. Once free of weight, switch to the high gear range.

Raise the landing gear all the way up. (Never drive with landing gear only part way up, as it may catch on railroad tracks or other obstructions.)

After raising the landing gear, secure the crank handle safely.

When the full weight of the trailer is resting on the tractor, do the following:

Check for enough clearance between the rear of the tractor frame and landing gear. (When the tractor turns sharply, it must not hit the landing gear.)

Check that there's enough clearance between the top of the tractor tires and the nose of the trailer.

Step 16: Remove and store wheel chocks in a safe place.

Uncoupling Tractor-Semitrailers

The following steps will help you to uncouple safely.

Step 1: Position rig.

Make sure the surface of the parking area can support the weight of the trailer. (Dirt, sand and gravel surfaces can cause problems, as can hot blacktop.)

Have the tractor lined up with the trailer. (Pulling out at an angle can damage landing gear.)

Step 2: Ease pressure on locking jaws.

Shut off trailer air supply to lock trailer brakes.

Ease pressure on the fifth wheel locking jaws by backing up gently (this will help you release the fifth wheel locking lever).

Put parking brakes on while the tractor is pushing against the kingpin. This will hold rig with pressure off the locking jaws.

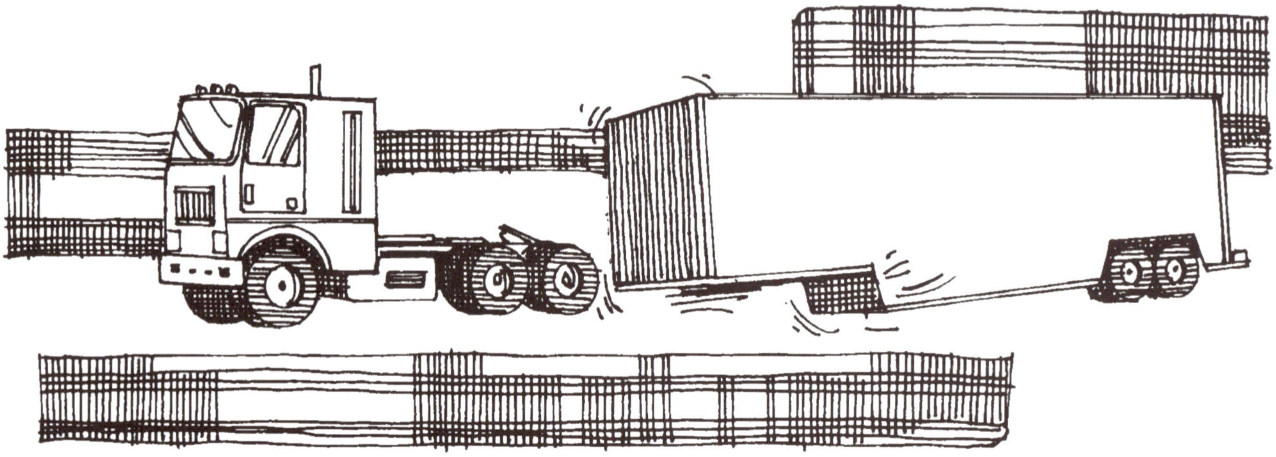

Step 3: Chock trailer wheels.

Chock the trailer wheels if the trailer doesn't have spring brakes, or if you're not sure. (The air could leak out of the trailer air tank, releasing its emergency brakes. The trailer could then move if it didn't have chocks.)

Step 4: Lower the landing gear.

If the trailer is empty -- lower the landing gear until it makes firm contact with the ground.

If trailer is loaded, after the landing gear makes firm contact with the ground, turn crank in low gear a few extra turns; this will lift some weight off the tractor. (Do not lift the trailer off the fifth wheel.) This will make it easier to unlatch the fifth wheel and to couple next time.

Step 5: Disconnect air lines and electrical cable.

Disconnect air lines from trailer. Connect air line glad hands to dummy couplers at back of cab, or couple them together.

Hang electrical cable with plug down to prevent moisture from entering it.

Make sure lines are supported so they won't be damaged while driving the tractor.

Step 6: Unlock the fifth wheel.

Raise release handle lock (if so equipped).

Pull the release handle to open position.

Keep legs and feet clear of the rear tractor wheels to avoid serious injury in case the vehicle moves.

Step 7: Pull tractor partially clear of trailer.

Pull the tractor forward until the fifth wheel comes out from under the trailer.

Stop with tractor frame under trailer (prevents trailer from falling to ground if landing gear should collapse or sink).

Step 8: Secure the tractor.

Apply parking brake.

Place transmission in neutral.

Step 9: Inspect trailer supports.

Make sure ground is supporting trailer.

Make sure landing gear isn't damaged.

Step 10: Pull tractor clear of trailer.

Release parking brakes.

Check the area, and drive the tractor clear.

Coupling Twin Trailers

Step 1: Secure the second (rear) trailer.

If the second trailer doesn't have spring brakes, drive the tractor close to the trailer, connect the emergency line, charge the trailer air tank, and disconnect the emergency line. This will set the trailer emergency brakes (if the slack adjusters are correctly adjusted). Chock the wheels.

Step 2: Couple the tractor and first semi-trailer as described earlier. For safe handling on the road, the more heavily loaded semi-trailer must always be in first position behind the tractor. The lighter trailer should be in the rear.

Step 3: Position the converter dolly in front of the second (rear) trailer.

Release dolly brakes by opening the air tank petcock. (Or, if the dolly has spring brakes, use the dolly parking brake control).

If distance is not too great, wheel the dolly into position by hand so it's in line with the kingpin.

Or, use tractor and first semi-trailer to pick up the converter dolly: Position combination as close as possible to converter dolly.

Move dolly to rear of first semi-trailer and couple it to the trailer.

Lock pintle hook.

Secure dolly support in raised position.

Pull dolly into position as close as possible to the nose of the second semitrailer.

Lower dolly support.

Unhook dolly from first trailer.

Wheel dolly into position in front of second trailer in line with the kingpin.

Step 4: Connect the converter dolly to the front trailer.

Back first semitrailer into position in front of dolly tongue.

Hook dolly to front trailer.

Lock pintle hook.

Secure converter gear support in raised position.

Step 5: Connect the converter dolly to the rear trailer.

Make sure trailer brakes are locked and/or the wheels chocked.

Make sure trailer height is correct. (It must be slightly lower than the center of the fifth wheel, so the trailer is raised slightly when dolly is pushed under.)

Back converter dolly under the rear trailer.

Raise landing gear slightly off the ground to prevent damage if the trailer moves.

Test coupling and pulling against pin of the number two semitrailer.

Make visual check of coupling. (No space between upper and lower fifth wheel; locking jaws closed on kingpin.)

Connect safety chains, air hoses and light cords.

Close converter dolly air tank petcock, and shut-off valves at rear of second trailer (service and emergency shut-offs).

Open shut-off valves at rear of first trailer (and on dolly if so equipped).

Raise landing gear completely on second trailer.

Charge trailers (push air supply knob in) and check for air at rear of second trailer by opening the emergency line shut-off. If air pressure isn't there, something is wrong and the brakes won't work.

Uncoupling Twin Trailers

Step 1: Uncouple rear trailer.

Park rig in a straight line on firm, level ground.

Apply parking brakes so rig won't move.

Chock wheels of second trailer if it doesn't have spring brakes.

Lower landing gear of second semitrailer enough to remove some weight from dolly.

Close air shut-offs at rear of first semitrailer (and on dolly if so equipped).

Disconnect all dolly air and electric lines and secure them.

Release dolly brakes.

Release converter dolly fifth wheel latch.

Slowly pull tractor, first semitrailer and dolly forward to pull dolly out from under rear semitrailer.

Step 2: Uncouple converter dolly.

Lower dolly landing gear.

Disconnect safety chains.

Apply converter gear spring brakes, or chock wheels.

Release pintle hook on first semitrailer. Caution: Never unlock the pintle hook with the dolly still under the rear trailer. The dolly tow bar may fly up, possibly causing injury, and, at the very least, making it very difficult to re-couple.

Slowly pull clear of dolly.

Other Combinations

The methods and procedures described here apply to the most common tractor-trailer combinations. However, there are additional types of combinations, as well as other ways of coupling and uncoupling them. While they won't be covered here, it's your responsibility to learn the right way to couple the vehicle(s) you'll drive according to the manufacturer and/or the owner.

coupling & uncoupling review

1. What might happen if the trailer is too high when you try to couple?

2. After coupling, how much space should be between the upper and lower fifth wheel?

3. You should look into the back of the fifth wheel to see if it's locked onto the kingpin: True or False?

4. To drive you need to raise the landing gear only until it just lifts off the pavement: True or False?

5. What is a converter dolly?

6. What are shut-off valves?

7. Where should the heaviest trailer be in a set of doubles?

8. Which shut-off valves should be open and which closed?

9. How can you test that air flows to all trailers?

10. How can you test the tractor protection valve?

11. How can you test the trailer emergency brakes?

12. How can you test the trailer service brakes?

Special Rigs 5 – 5

multiple articulation vehicles

oversized vehicles

low clearance vehicles

high center of gravity vehicles

unstable loads

refrigerated trailers

special cargo vehicles

special handling vehicles

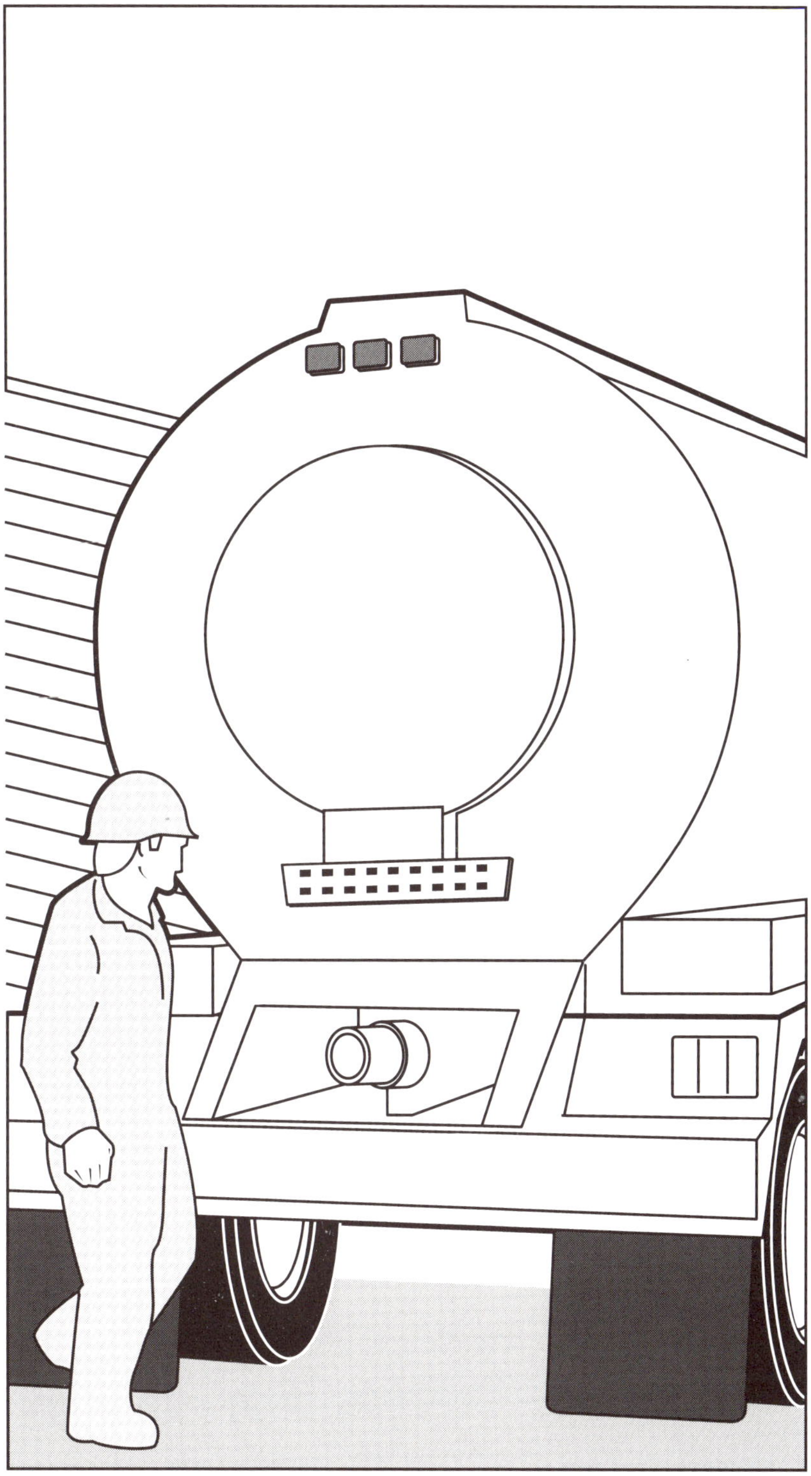

Special Rigs 5 – 5

A special rig is any combination vehicle that differs from the standard tractor and 48-53 foot dry freight van trailer with five axles and 18 wheels.

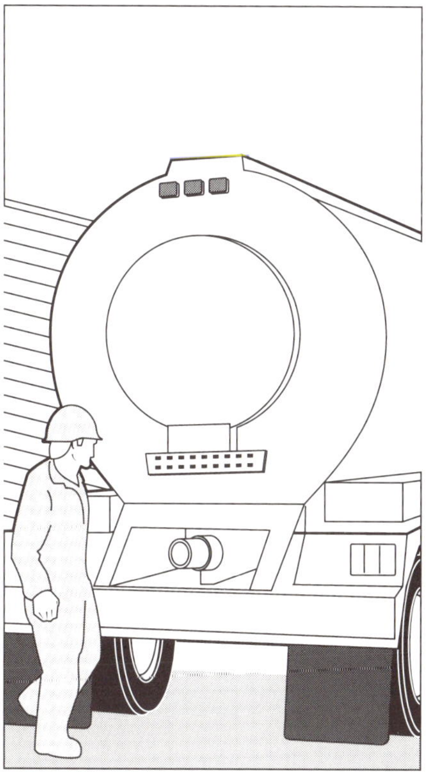

This definition includes the following types of vehicles:

Vehicles with more than one point of articulation (joints);

Vehicles that are over length, over height, over width or over weight;

Vehicles with very low ground clearance;

Vehicles that, when loaded, have a high center of gravity; Vehicles with load stability problems;

Vehicles used for certain special cargos;

Vehicles that require special handling.

Most of the material covered in this manual is applicable to the operation of all tractor-trailer combinations, but special rigs can require special procedures as well. (Federal and state regulations often restrict operation of special rigs in some way. These restrictions should be well known to companies that use special rigs.)

Multiple Articulation Vehicles

Common multiple articulation vehicles include twin trailers, triples, and B-trains. The two basic types of twin trailers are western doubles and turnpike doubles.

western doubles

Western doubles are made up of two semi-trailers, with the second trailer converted to a full trailer using a converter gear (set of wheels with a fifth wheel). The converter gear is connected to the first trailer by an eye with a pintle hook on the back of the first trailer.

While a standard combination has one point of articulation, or joint, doubles have three: First trailer kingpin and fifth wheel; pintle hook and eye; converter gear fifth wheel and kingpin of

experience speaking

"You have to make sure that you load the trailers and put them in their proper perspective. In other words, don't get one heavy one in the back end and one light one in the front end. You want the weight on your pulling vehicle."

Robert
Instructor
Missouri

the second trailer. Western double trailer length is typically 24-31 feet, and overall length comes out to 60-75 feet. The trailers may be dry vans, tankers, flatbeds or dumps.

There are several notes concerning special handling when pulling western doubles. Avoid backing. This typically difficult maneuver is made even more difficult with a double trailer unit not designed to be backed. Smooth steering is required. Jerking or whipping of the steering wheel is greatly magnified in the second trailer. (To create a picture in your own mind, think of a group of ice skaters playing crack the whip.)

The heaviest trailer must always be ahead of the lighter one. Don't apply brakes in a curve, because this will cause the second trailer to dip. The greater length requires the driver to take greater caution in passing, lane changing and crossing intersections. It's impossible to make tight turns with close coupled rigs. Bumps and potholes may cause the tops of the trailers to strike each other. On curves, the driver needs to be aware of tracking to stay within the right lane while going around.

Finally, doubles add steps to the normal coupling and inspection procedures. Drawbar and pintle hook must be properly in place and hooked. Safety chains need to be in place. Light cords need to be attached. All airlines must be hooked up properly and valves opened or shut as required. Additional information on coupling and uncoupling was presented in the previous section (5-4 "Coupling And Uncoupling").

turnpike doubles

The main difference between western and turnpike doubles is size. Trailer length is 35-40 feet, with overall length around 100 feet. Typically a high output engine and multiple gear range transmission are required.

Other major characteristics, handling and special requirements are essentially the same as for western doubles, if magnified somewhat by the additional length. However, a driver who pulls turnpike doubles does need to be aware of areas where their use is restricted, as well as special permits that are required. A company that regularly uses turnpike doubles should be aware of current federal and state regulations concerning their operation.

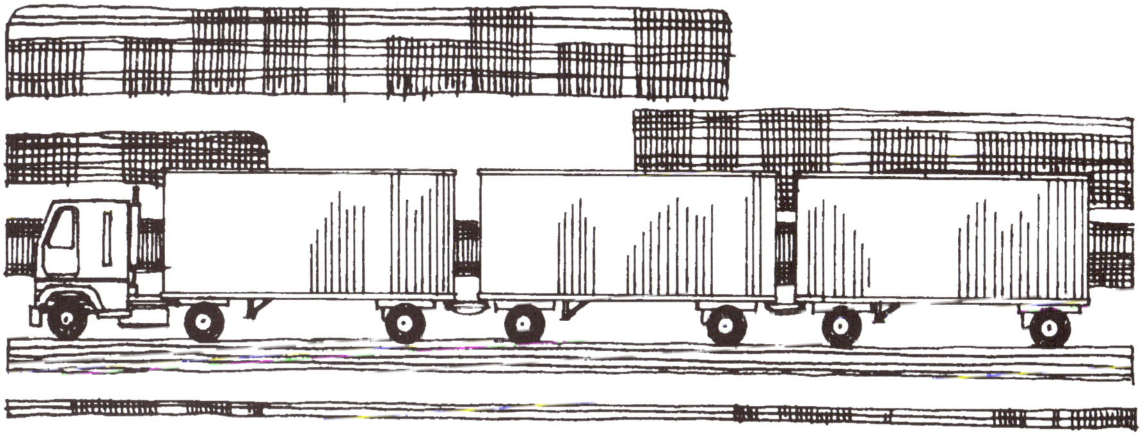

triple trailers

Triples are becoming a much more common sight on America's highways. Triples are made up of three semi-trailers, with the second and third converted to full trailers using converter gear and connected, as with doubles, by drawbar and pintle hook. There are three kingpin fifth wheel connections and two eye-pintle hook connections. Trailers are typically 27 feet long.

Most characteristics and handling are the same as doubles with the obvious exaggeration for length and number of trailers. Areas and highways where triples are allowed are limited. This and other regulatory information should be kept current by any company using triples in its operations.

b-trains

B-trains are made up of two semi-trailers. On the first trailer, tandem rear axles are to the rear of the trailer body so the second axle is under the nose of the second semi-trailer. A fifth wheel mounted above the second axle eliminates the need for a converter gear. This eliminates one of the points of articulation of a regular double.

Trailer and overall length will vary depending on the geographical locations of B-train operations. Generally there will be two 40-foot trailers or one 40-foot and one 27-foot trailer. B-trains are more common in Canada than in the United States.

Oversize Vehicles

Oversize vehicle trailer types may include lowbeds, drop frames, flatbeds and open top vans. There can be a number of wheel and axle combinations depending on weight, size of cargo and state regulations. Many oversize vehicles are equipped with outriggers to support oversized loads. Converter gears may be attached to the trailer or tractor by fifth wheel and kingpin, or to the cargo itself to distribute weight over more axles and support longer loads.

Typically, the cargo that creates or makes necessary an oversize vehicle is over dimensional or overweight. This could include such cargo as power plants, industrial dryers or heavy construction equipment.

Oversize vehicles are usually operated with special permits and on designated highways. Requirements will vary depending on type of cargo, size, weight and state regulations.

Low Clearance Vehicles

There are two basic kinds of low ground clearance vehicles, double drop frame and drop frame (flatbed). A double drop frame drops between the kingpin and trailer axles. A drop frame drops right behind the kingpin. Low clearance vehicles are designed to haul heavy, oversized cargo or to accommodate greater cubic capacity loads. Operation of low clearance vehicles sometimes requires special licenses and permits, as well as restricted road use, depending on trailer type, size, weight of cargo and individual state requirements.

High Center of Gravity Vehicles

The greatest danger in operating high center of gravity vehicles is their tendency to tip. There are generally few regulatory restrictions, but great care is required in vehicle handling.

Examples of high center of gravity vehicles include dry bulk tankers, livestock transports, reefers (refrigerated units) or some oversized vehicles.

Unstable Loads

The two most common unstable load vehicles are liquid tankers and trailers designed for hauling livestock.

liquid tankers

Liquid tankers are generally semi-trailers either oval, round or occasionally even square in shape. There are cold tankers, hot tankers and pressurized tankers. Often tankers are used to transport hazardous materials.

Liquid tankers vary in length. They may contain one compartment or several. They may or may not contain baffles. Handling can be difficult to master because of surging loads that make for an unstable vehicle.

Examples of liquid tankers include petroleum or chemical tankers, acid tanks, liquefied gas tanks, insulated tanks and milk tankers.

Some specific training is necessary for tanker drivers depending on the type of trailer and cargo. Road use may be restricted and permits required, also based on the type of trailer and cargo.

Driver inspections must include checking for leaks, and checking all hoses, valves, fittings and the emergency valve release. Thereafter, the standard inspection procedures apply.

livestock transport

Livestock transport includes hauling live cattle, hogs, horses, sheep and a variety of other less common living freight. A semi-trailer with a flat floor or double drop frame is used. Slots or holes in the sides allow livestock to

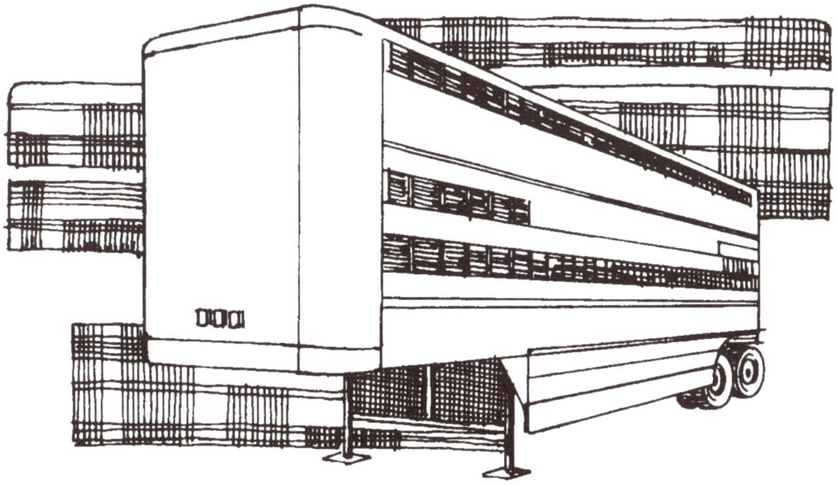

breathe. Livestock trailers often have side doors in addition to, or instead of, rear doors.

Trailer length will vary from 27 to 45 feet. Trailers will have a fixed tandem axle. Smaller animals such as pigs or sheep are often carried in trailers converted to accommodate two or three decks.

Live cargo can create problems. Put simply, animals shift. You need to maintain a speed where you can safely keep the vehicle under control. When stopping, tap brakes lightly to set the animals, then gradually apply brakes. Never attempt to haul livestock without proper training. Livestock hauling is unique in that the driver is responsible for the passengers' health and safety en route.

Refrigerated Trailers

Refrigerated trailers, or reefers, come in basically two types. On nose mounts, the refrigeration unit is mounted on the upper front of the trailer. On belly mounts, the refrigeration unit is mounted under the trailer.

Reefers are most often van, or box, type trailers. Some have racks or rails suspended from the roof, on which animal carcasses are hung, and which can make them unstable.

Others have separate compartments to keep some cargo frozen while other cargo is kept cool. Slotted floors allow for air or gas circulation. Floors, walls and roof are insulated.

Vegetables, fruit, meat and other food stuffs are common reefer cargo, as are some chemicals.

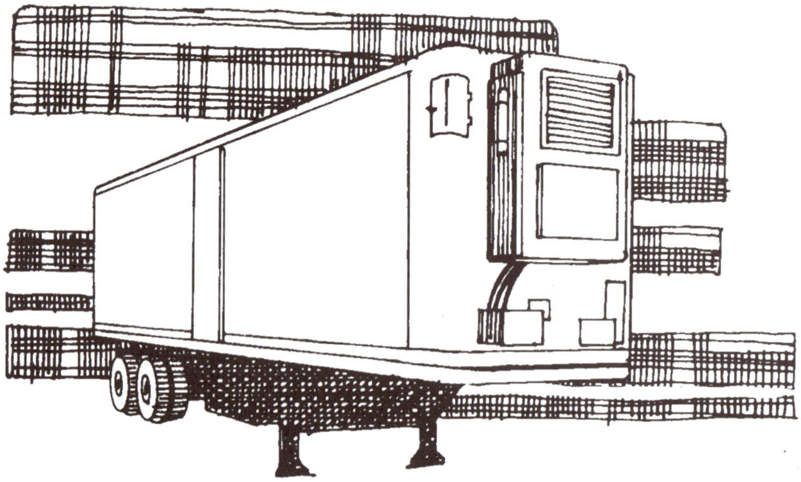

Refrigerated trailers have their own engines powered by gasoline, diesel fuel or liquefied petroleum gas. Obviously, they also have their own fuel tank.

Special reefer inspection procedures include checking for holes in walls, checking ceiling and floor ducts, and door and door gasket inspection. Fuel, coolant, oil and refrigerant levels must also be checked regularly.

Special Cargo Vehicles

A special cargo vehicle is any vehicle designed to haul primarily one particular type of cargo. Two of the most common types of special cargo vehicles are pole trailers and auto transporters.

Special Handling Vehicles

Some vehicles present special handling problems, because of visibility, location of power plant or location of the steering axle. A number of city pickup and delivery tractors are special handling vehicles.

In snub-nose tractors, the rear of the engine protrudes into the cab. They're often used for close clearance city work. Half cabs have a heavy duty diesel engine mounted next to the cab. They're most frequently used as yard horses, or goats. Dromedary tractors have a cargo body mounted between the rear of the cab and the fifth wheel.

special rigs review

1. What is a special rig?

2. Name five vehicles or combinations that qualify as special rigs.

3. What are some common characteristics of refrigerated trailers?

4. What do liquid tankers and livestock trailers have in common?

5. What is one common problem in operating a high center of gravity vehicle?

6. What's the difference between western doubles and turnpike doubles?

Basic Control: Proficiency Development 6

Seeing 6 – 1

scanning

mirrors

seeing to the rear

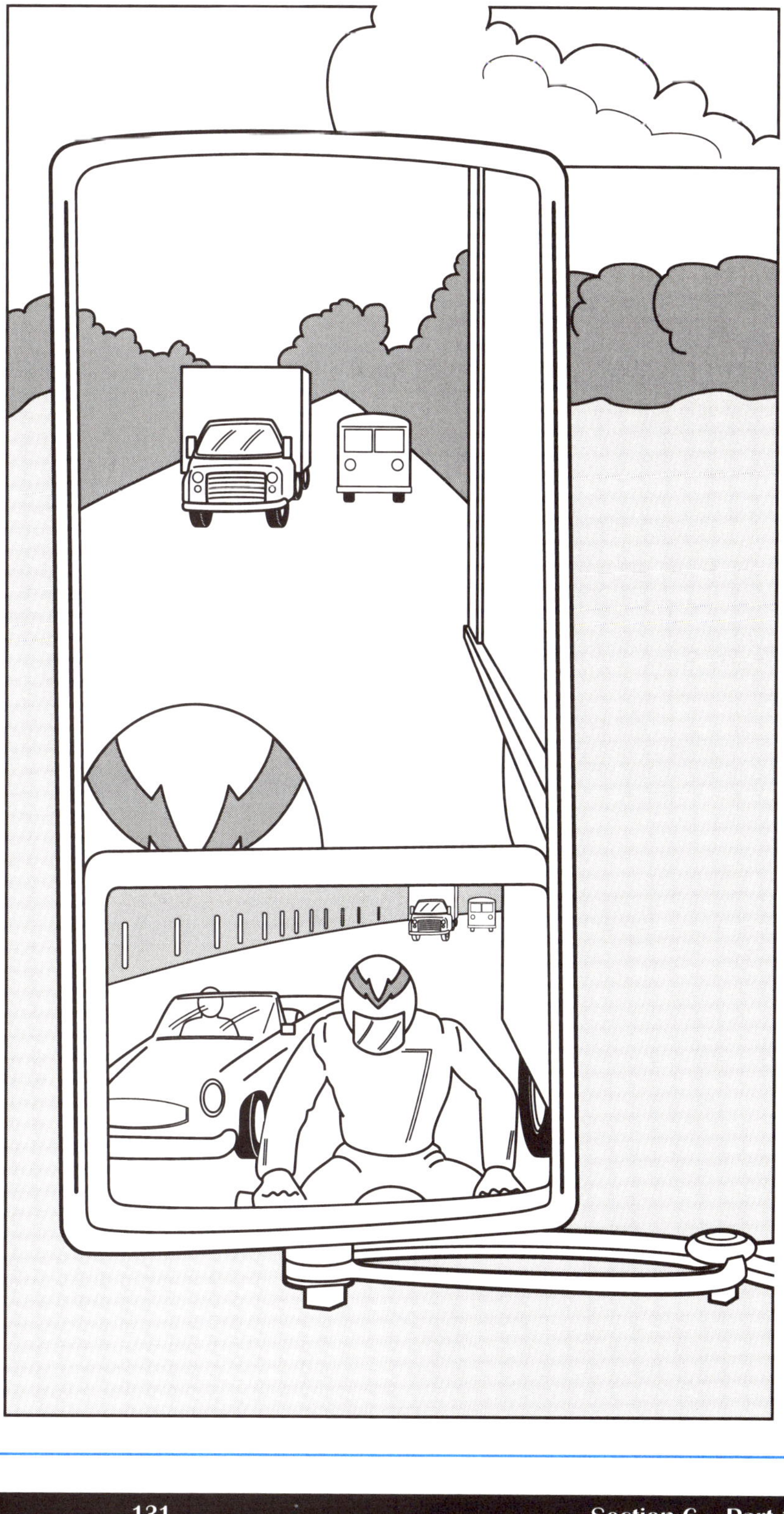

Seeing 6 – 1

To be a safe driver you need to be aware of what's going on all around you and your vehicle.

Not looking and seeing properly is a major cause of accidents. It is inevitable that all drivers look ahead. But many do not look far enough ahead. And that is to say nothing about keeping a view to the sides and the rear.

There are several common seeing errors that cause accidents. Perhaps the driver's view is blocked or limited in some way. Or the driver may simply be looking in the wrong place at the wrong time. Long rides tend to leave you driving with a fixed stare. Sometimes an accident is caused for no reason other than that the windshield and mirrors are not clean and clear.

In some ways, the view from a truck tractor seat can be both the best and the worst. The additional height you have over other drivers allows you to see over other traffic, providing a better view of the action ahead. On the other hand, a driver's normal position in the cab causes difficulty in seeing to the sides and rear. Small vehicles are often hard to see, as is any vehicle on the right side of you or positioned along side the drive wheels on either side.

Scanning

A good driver continually scans his entire sight area. He looks far ahead. He looks at the vehicle ahead of him and the one ahead of that. He keeps an eye on vehicles to both the left and right. He constantly monitors vehicles behind him. The good driver realizes that he must make the most of what he sees through the windshield and windows, as well as in his rearview mirrors. He physically turns to see either side.

distance scanning

Looking way ahead helps you to drive smoothly, efficiently and safely. Always aim your vehicle at an imaginary "target" at the center of your line of travel and as far ahead as possible. Look far enough ahead to give yourself time to spot a problem, decide the best way to avoid it, check for traffic conditions that might keep you from making the best move and, finally, to maneuver out of a potential problem situation.

Look to the sides as well as the center of the road. Distance scanning allows for a wider scene to be viewed.

As a general rule, you should look 10 to 12 seconds ahead of your vehicle; depending on the amount of traffic and the available sight distance. In the city, 12 seconds equals about one average block. On the highway, 12 seconds can equal a quarter mile or more. If you can't see that far ahead, you should tell yourself to slow down.

It would be a mistake to look off into the distance at all times. You need to look around. Check the status of vehicles beside and behind you. Check your vehicle's instruments. Keep an eye on road signs. Keep your eyes moving from far to near.

Distance scanning yields several benefits. Safety is increased by identifying hazards early. Fuel and time are saved by spotting traffic situations early. The driver of a tractor-trailer needs more time to react to situations than the driver of an automobile. Distance scanning gives you the time, and the opportunity, to avoid abrupt stops and radical speed adjustments. You can adjust smoothly to traffic and situations.

scanning to the sides

While driving, you should periodically scan to the sides. You must look to the sides in certain critical situations.

At crosswalks, pay particular attention to the right. Pedestrians, bicycles and mopeds, for example, are often out of sight when closest to your vehicle. Yield the right of way to pedestrians when turning on green. At intersections learn to look left, right, left and continue scanning as you pull through. At all times, watch out for blind spots created by your mirrors or other obstacles.

There are also several other typically busy places where it's critical that you scan to the sides. In school zones, watch for children darting into traffic. At construction sites, vehicles and workers will seem to be everywhere, moving in all directions at once. At malls and shopping centers, vehicles pull into the flow of traffic and drive against traffic, while pedestrians appear from around corners and behind parked vehicles. Near hospital zones or fire stations, emergency vehicles appear suddenly, sometimes with little, if any, warning.

You can see better pulling away from intersections if you nose your vehicle out very slowly. Stop twice: First at the stop sign or light, then again when you have eased forward to where you can see oncoming traffic. Give drivers on the cross street a chance to see you and yourself a chance to see them.

Mirrors

The importance of using your rearview mirrors while driving can't be underestimated. The mirrors provide your only view to the rear of your vehicle. It's crucial that you check your mirrors before changing speed or position in traffic. A good "rule of thumb" is to check your mirrors every four seconds. It's especially important that you use your mirrors (in addition to glancing around) to check your blind spots (alongside your tractor's fuel tank and drive wheels).

There are two types of mirrors, plane and convex. Most tractors are equipped with both. A plane mirror alone allows too large a blind spot. A convex mirror gives a distorted view.

A plane mirror gives you the best view of the rear of your trailer and the roadway behind. It doesn't give a wide view, and it leaves a blind area along most the length of the vehicle. Your field of view is greatest in the left mirror, simply because the closer the mirror, the larger the image; the larger the image, the greater your field of view.

There are blind spots on both sides, making lane changes and other maneuvers risky. Tight turns are also a problem, because you can't detect adjacent small vehicles or pedestrians. While using your mirrors to their fullest advantage, you must still signal and wait a moment before changing direction.

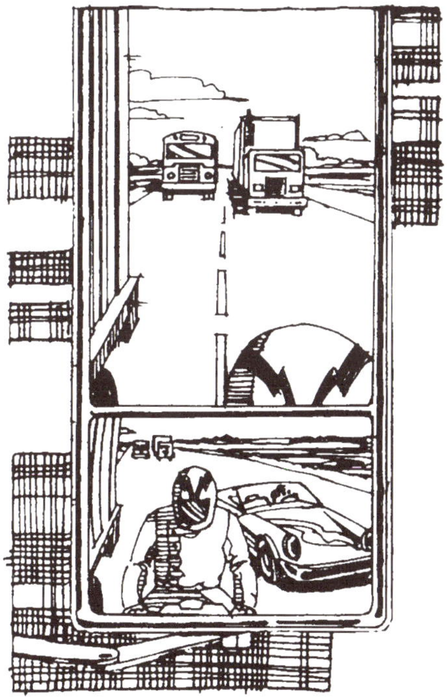

Images in your side mirror will appear similar to those in the side mirrors of a car. You should easily learn to judge the speed and distance of overtaking vehicles using your plane mirror.

Convex mirrors are designed with an outward curvature to provide a wide angle view. They give a broader view than plane mirrors and eliminate much of the blind area. They provide the best close-up view of the sides of your tractor-trailer.

Unfortunately, however, convex mirrors show a distorted image. Overtaking vehicles will appear smaller and farther away than they actually are. The small, stick-on mirrors many drivers use give the most distortion, while reducing the plane mirror viewing area.

A combination of plane and convex mirrors is best. This will provide both maximum side and rear vision. But the combination does take a little getting used to. The overlapping field of view can confuse the driver. It helps to have a high quality, properly adjusted and aligned convex mirror to reduce overlap and distortion. Do note, also, that even with the right combination of mirrors, small blind spots will remain. Use your mirrors continually.

Get used to their position and the field of view each mirror gives you. Don't ever move your vehicle until your mirrors are properly positioned. Frequent checking of mirrors is a necessity.

Take care of your mirrors. Adjust them until they allow the greatest and best view possible, before pulling out. Keep your mirrors clean and tight to make them the most effective under all conditions, including glare, extreme weather conditions and where dirt and mud are a problem. If your mirrors aren't tight, they will vibrate and distort your rear and side views.

Seeing To the Rear

As you drive, you need to continually use your mirrors to monitor to the rear of your vehicle. There are a number of things you should be looking at and for. Check load and cargo security, watch for loose or falling cargo. Check for tire fires. Check for vehicles adjacent to your rig. Know what's beside the tractor or trailer. Know what's happening and what has happened beside and behind you.

You need to use your mirrors when changing direction or when forced to slow down quickly or unexpectedly. Lane changes, merging and turns all require mirror checks, as do alleys, intersections, traffic tie-ups and while approaching or driving alongside parked or stopped vehicles. Special situations require more than regular mirror checks.

Going back to where we began, to be a safe driver, you need to know what's going on ahead of and all around your vehicle. The importance of seeing can't be overestimated.

experience speaking

"The basic problem we've had with semis is for them to take extreme care in making wide turns. We post our vehicles that they will in fact do that. However, we still must be in a defensive mode, and once they've executed the turn, we tell them to watch that passenger mirror very closely to make sure that no small objects have appeared in there that they will run over."

John
Safety Manager
Tennessee

seeing review

1. How far ahead should you look?

2. What are two main things to look for ahead?

3. What's your most important and most effective way to see to the sides and rear?

4. Not looking and seeing properly is a major cause of accidents: True or False?

5. What are some common seeing errors that cause accidents?

6. What is meant by scanning?

7. What part do mirrors play in seeing the whole picture?

Communicating 6 – 2

communicating intent

communicating presence

communications from others

Communicating 6 – 2

Closely related to the issue of seeing, is the issue of communicating.

In fact, many of the messages you communicate to others and that others communicate to you will be visible communication; messages that are seen. Of course, there are also audible communications, involving, for the most part, the proper use of vehicle horns. There would be far fewer accidents if all drivers were excellent communicators and excellent observers.

Communicating Intent

Other drivers can't know what you're going to do if you don't tell them. Signaling your intentions is important to safety. You have four basic methods of signalling, or visibly communicating your intent: Use of turn signals; warning flashers; flashing brake lights; and altering the position of your vehicle.

turn signals

Always communicate your intent to change direction. Other drivers expect you to continue in the same direction if you don't let them know when you intend to do something different. Remember, too, that the most dangerous vehicles are the ones you don't see, so signal your intent even when it appears that no one is around you. Signal anytime you plan to turn, change lanes, pass another vehicle, merge, parallel park or pull into traffic from the curb. Some good rules for signalling follow.

Signal early. Signal well in advance of turning or changing lanes. It's the safest way to keep others from trying to pass you. Give others enough time to receive and react to your communication.

In city traffic, signal one-half block beforehand. On the open highway, 500 feet is a good rule. It's illegal to make turns without signalling first. Do not, on the other hand, signal too early, because you will confuse other drivers.

Signal continuously. You need both hands on the wheel to turn safely. Don't cancel the signal until your maneuver is complete.

After you've completed a turn or lane change, cancel your

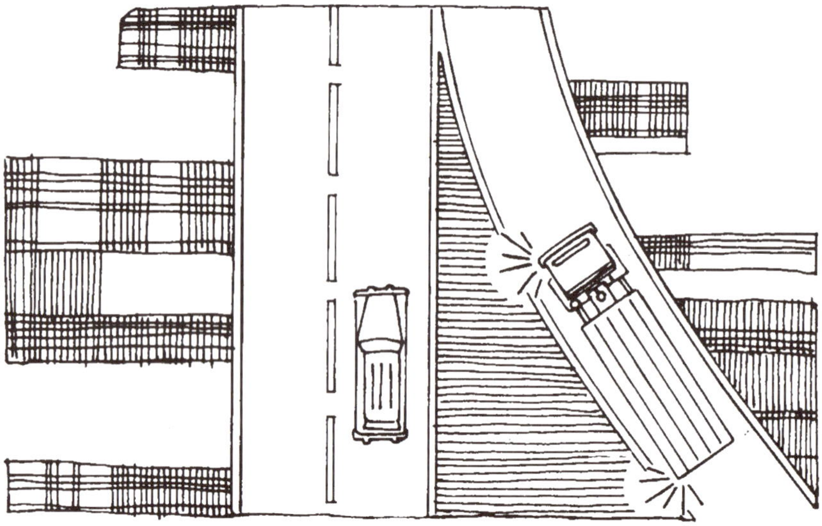

signal. Even if you have self-cancelling signals, check to make sure that they've worked. Cancelling your signal before shifting establishes an association between cancelling and shifting, and helps you develop a cancelling habit.

Make no turns or lane changes without being certain that others have seen your signals and you know that it's safe.

slowing down

Warn drivers behind you when you need to slow down. A few light taps on the brake pedal — enough to flash the brake lights — should warn following drivers. Use the four-way emergency flashers for times when you're driving very slow or are stopped.

Keep in mind the size of your vehicle. It's difficult for fellow drivers to see ahead of large tractor-trailers.

This makes it particularly important that you communicate when traffic necessitates slowing or stopping. There are any number of situations when other drivers don't expect you to slow down or stop. Slowing or stopping in the middle of a block or even on the open highway is common. Many times there will be such things as stalled traffic or objects in the road which you can see, but drivers behind you can't.

One thing you should never do is direct traffic. Some drivers try to help others out by signalling when it's safe to pass. You may be responsible for an accident, which in turn could result in injury to yourself and others, in addition to thousands of dollars worth of damage.

Communicating Presence

Other drivers may not notice your vehicle even when it's in plain sight. Let them know you're there to help prevent accidents.

Whenever you're about to pass a vehicle, pedestrian or bicyclist, assume they don't see you. They could suddenly move in front of you. When it's legal and safe, tap the horn lightly or, at night, flash your lights from low to high beam and back (only when safe and legal). Drive carefully enough to avoid a crash even if others don't see or hear you.

At dawn or dusk or in rain or snow, you need to make yourself easier to see. If you're having trouble seeing other vehicles, other drivers will have trouble seeing you. Turn on your lights. It's law in some states if wipers have to be on, headlights must be on.

Use the headlights, not just the identification or clearance lights. Use the low beams; high beams can bother people in the daytime just as at night.

When you pull off the road and stop, be sure to turn on the four-way emergency flashers. This is especially important at night. Don't trust the taillights to give warning. Drivers have crashed into the rear of a parked vehicle many times because they thought it was moving normally.

experience speaking

"Actually, what we want to do is give the traffic around us as much notice as we can. We want to get that directional signal going, and we might as well start reducing the speed when we recognize that it is a safe time to do so. On many of the exit ramps you have plenty of time before you actually enter into that exit ramp."

John
Instructor
Illinois

warning devices

You should put out your reflective triangles within 10 minutes if you must stop on the road or shoulder of the road for an extended period. Place your warning devices on the traffic side of the vehicle, within 10 feet of the front or rear corners to mark the location of your vehicle. Also place warning devices about 100 feet behind and ahead of your vehicle, on the shoulder or in the lane you're stopped in. If there's any hill, curve or other obstruction that will prevent other drivers from seeing your vehicle within 500 feet, place a triangle beyond that obstruction. If you must stop on or by a one-way or divided highway, place warning devices 10 feet, 100 feet and 200 feet toward the approaching traffic.

When putting out triangles, hold them between yourself and oncoming traffic for your own safety.

It can't be stressed too heavily or repeated too often that many accidents have occurred when drivers, not realizing a vehicle was stopped, drove right into it!

NOTE: WARNING DEVICE PLACEMENT will be covered with greater depth and detail under the heading "Accident Procedures" (7-10).

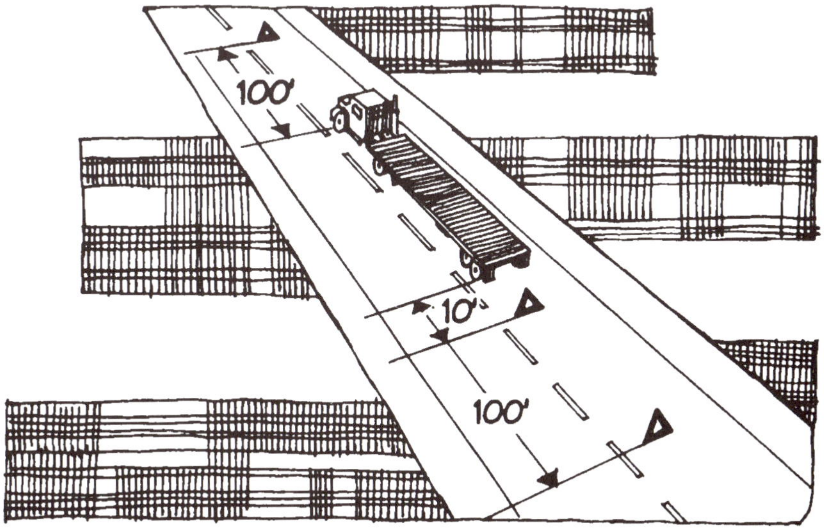

using your horn

Your horn lets others know you're there. It can help to avoid an accident. Use your horn only when it is needed. Remember, your horn can startle others and can be dangerous when used unnecessarily. In general, limit horn use to your electric horn. An air horn is extremely loud. It can frighten other road users.

Use your air horn in cases of immediate danger only. Don't socialize with your horn. Your horn should be used for two purposes: Communicating your presence to avoid danger and to warn of immediate danger.

cb radio

The CB can be a good communication tool if used safely and legally. The CB can also be distracting and dangerous if improperly used. CBs can be used to provide or obtain information about weather and traffic conditions. They can come in very handy at the scene of an accident or other emergency. Finally CBs can be legitimately used to obtain directions or other helpful local information.

Idle chatter, which distracts the driver and ties up one hand that could be better put to use controlling the vehicle, is an unsafe use of the CB. To use profanity over the CB or to use it to inform others that it's safe to speed, are both illegal.

Communications From Others

It isn't enough to communicate well yourself on the road. You also have to be alert to the intentional and unintentional communications you receive from others.

Besides the intentional communication of turn signals, lights and vehicle horns, there are other, more subtle communications from other drivers or vehicles the good tractor-trailer driver learns to recognize.

Driver movement in a vehicle, such as shifting around in the seat or looking in the mirror, may indicate possible directional changes. Eye contact between drivers or drivers and pedestrians reduces the chance of sudden, unexpected maneuvers. Sometimes, however, eye contact will make others think you plan to yield when you expect them to.

Some communications are received by observing other vehicles. Sudden slowing may indicate a direction change. A slight position change within the lane could be the same message, regardless if the turn signals are on or not. Turned front wheels or visible exhaust from a parked car should tell you that vehicle is going to pull out.

Other drivers may signal too late or too soon. They might not signal at all. They may fail to cancel a signal. It's up to you to be ready for anything by reading all kinds of communication as best as you can.

The professional driver is alert to the communications of others. He anticipates and avoids conflict.

communicating review

1. What does communicating mean in safe driving?
2. Where should you reflective triangles be placed when stopped on a divided highway?
3. Name some ways of communicating intent.
4. When is it not appropriate to use your horn?
5. What might it mean if a driver behind you is driving erratically: Slowing down, speeding up, changing lanes?
6. What are two ways of communicating your presence?

Managing Speed 6 – 3

stopping distance

obeying the speed limit

Managing Speed 6 – 3

Driving too fast is a major cause of fatal crashes.

Managing speed means adjusting to road conditions, curves, visibility, traffic and hills. You must adjust your speed to specific driving conditions such as traction, curves, visibility, traffic and hills. Much of that adjustment involves the ability to stop.

Stopping Distance

There are three elements that add up to total stopping distance:

	Perception Distance
+	Reaction Distance
+	Braking Distance
=	Total Stopping Distance

Perception distance is the distance your vehicle travels from the time your eyes see a hazard until your brain recognizes it. The perception time for an alert driver is about 3/4 second. At 55 mph, you travel 60 feet in 3/4 second.

Reaction distance is the distance traveled from the time your brain tells your foot to move from the accelerator until your foot is actually pushing the brake pedal. The average driver has a reaction time of 3/4 second. This accounts for an additional 60 feet traveled at 55 mph.

Braking distance is the distance it takes to stop once the brakes are applied. At 55 mph on dry pavement with good brakes, it can take a heavy vehicle about 170 feet to stop. It takes about 4-1/2 seconds.

At 55 mph it will take about six seconds to stop, and your vehicle will travel about the distance of a football field. This is your total stopping distance (60 + 60 + 170 = 290 feet).

Whenever you double your speed, it takes about four times as much distance to stop, and your vehicle will have four times the destructive power if it crashes. High speeds increase stopping distances greatly. By slowing down a little, you can gain a lot in reduced braking, or stopping, distance.

The heavier the vehicle, the more work the brakes must do to stop it and the more heat they absorb. But the brakes, tires, springs, and shock absorbers on heavy vehicles are designed to work best when the vehicle is fully loaded. Empty trucks require greater stopping distances, because an empty vehicle has less traction. It can bounce and lock up its wheels, giving much poorer braking.

road surface

You can't steer or brake a vehicle unless you have traction. Traction is friction between the tires and the road. There are some road conditions that reduce traction and call for lower speeds.

It will take longer to stop, and it will be harder to turn without skidding, when the road is slippery. You must drive slower to be able to stop in the same distance as on a dry road. Wet roads can double stopping distance. Reduce your speed by about 1/3 (e.g., slow from 55 to about 35 mph) on a wet road. On packed snow, reduce your speed by half, or more.

If the surface is icy, reduce your speed to a crawl, and stop driving as soon as you can safely do so.

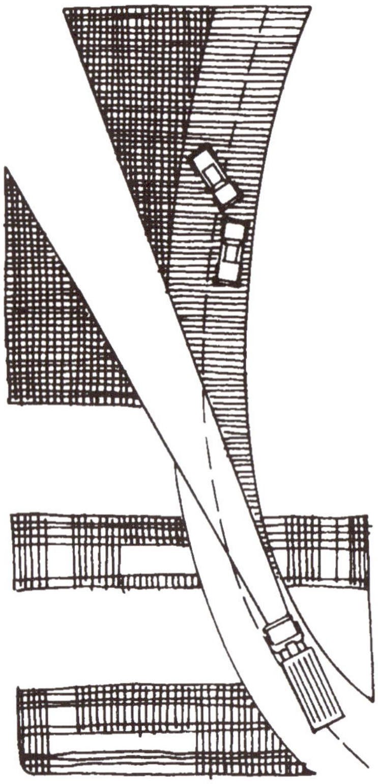

curves

Keep in mind that speed limits are designed with cars in mind. Drivers must adjust their speed for curves in the road. If you take a curve too fast, two things can happen. The wheels can lose their traction and continue straight ahead, so you skid off the road. Or, the wheels may keep their traction and the vehicle rolls over. Tests have shown that trucks with a high center of gravity can roll over at the posted speed limit for a curve. Many tractor-trailer drivers can relate this same information from their own unfortunate experience. A large rig will not, under any circumstances, handle as well on curves as an automobile.

Slow to a safe speed before you enter a curve, about 5 mph below the posted speed. Braking in a curve is dangerous because it's easier to lock the wheels and cause a skid. Slow down as needed. Do not exceed the posted speed limit for the curve. Be in a gear that will let you accelerate slightly in the curve. This will help you keep control.

visibility

You should always be able to stop within the distance you can see ahead. Fog, rain or other conditions may require that you slow down to be able to stop in the distance you can see. At night, you can't see as far with low beams as you can with high beams. When you must use low beams, slow down.

traffic flow

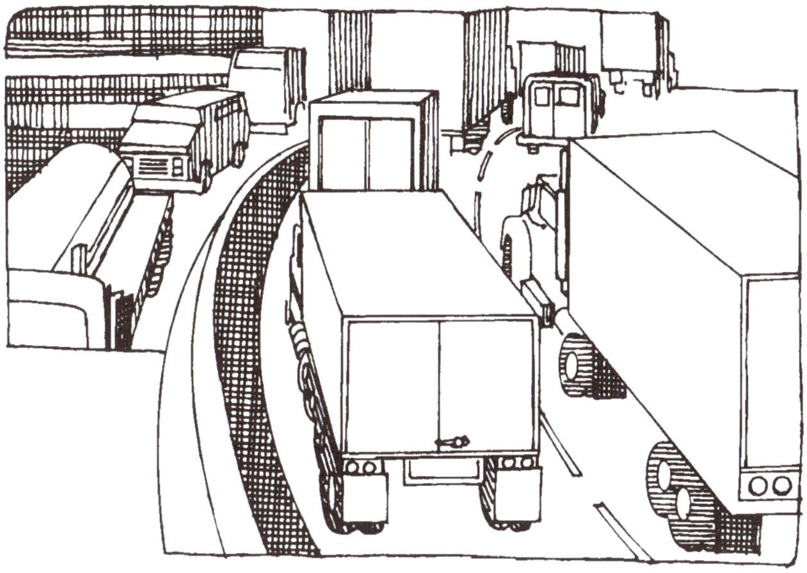

When you're driving in heavy traffic, the safest speed is generally the speed of other vehicles. Vehicles going the same direction at the same speed are not likely to run into one another. Drive at the speed of the traffic, if you can, without going at an illegal or unsafe speed. Keep a safe following distance. In heavy bumper-to-bumper city traffic, drive slightly slower than the traffic flow to maintain following distance.

Drivers exceed speed limits to save time. But anyone trying to drive faster than the speed of traffic will not be able to save much time. The risks involved aren't worth it. If you go faster than the speed of other traffic, you will have to keep passing other vehicles. This increases the chance of a crash. And it's more tiring. Fatigue increases the chance of a crash. Going with the flow of traffic is safer, and easier.

downgrades

Going slow is the most important thing in going down long steep hills safely. If you don't go slowly enough, your brakes can become so hot they won't slow you down. Shift your transmission to a low gear before starting down the grade. Pay attention to signs warning of long downhill grades, and check your brakes before starting down the hill. Use intermittent and moderate pressure on the brake pedal.

NOTE: UPHILL/DOWNHILL DRIVING will be covered with greater depth and detail under the heading "Extreme Driving Conditions" (6-6).

experience speaking

"The primary concern that we have with our people is that they blend with traffic, that they flow with the traffic situation. We do not want them to be in a position where they are going significantly slower than surrounding traffic, nor do we want them going significantly faster."

Gilbert
Safety Director
Georgia

Obeying The Speed Limit

Speed limits are based on the principles of good speed management. They take into account sight distance limitations, reduced traction, space limitations, previous accidents, type of vehicle, available light and fuel economy.

accidents

There are several good reasons not to speed. First and foremost is safety. The fact is that accidents are more likely at higher speeds. And studies show that the higher the speed a vehicle is traveling when there's an accident, the greater the chance of fatalities. The possibility of being responsible for the injury or death of someone else should be enough to slow anyone down to the speed limit. Unfortunately, that hasn't always been the case.

other penalties

Accidents and physical injuries, while they are the best reasons not to speed, are not the only risks you will take if you exceed the speed limit. Speeding also carries with it the possibilities of penalties.

For instance, there are traffic tickets. Tickets carry with them the obvious consequences of fines and lost time (speeding does not save time), but also some others that may be even more painful to take. You may lose your license. Your insurance rates may go up. You could lose your job!

maintenance costs

Higher speeds affect maintenance costs. Tires and brakes wear out faster at higher speeds. As your speed increases, so does engine heat and overall wear.

One study shows that maintenance costs increase 38% as speed increases from 50 to 60 mph, and 80% as speed increases from 50 mph to 70 mph. It's important to drive the speed limit to keep all costs down.

fatigue

Another cost of exceeding the speed limit falls in the area of driver fatigue. Continually trying to pass others, generally fighting traffic and constantly watching for police will wear you out. Add to this the ever-present noise and vibration of a tractor-trailer, which also increases at higher speeds, and the personal cost increases.

fuel economy

High — and rising — fuel costs make speed management vital. The ideal freeway speed for today's tractor-trailers is 50-55 mph. The power and fuel required to increase speed multiplies sharply: 73% more horsepower is needed to cruise at 60 mph; 159% more horsepower is needed to cruise at 70 mph.

poor public image

Many drivers and others in the trucking industry feel the public has the wrong impression of trucking and drivers. But often it's the speeding and tailgating tractor-trailer driver who is responsible for creating this image. Right or wrong, each time another motorist encounters a speeding truck driver, it reflects on the entire industry.

Remember, too, that irate citizens can report a driver to his employer and to law enforcement officials. Drivers are easily identified by vehicle numbers, company names, time and location. Such a report could result in the loss of your job.

why drivers speed

The two main reasons why tractor-trailer drivers speed are to shorten trip time and to make up for lost time. Tests show that time savings by speeding are rarely significant. At best, you might save a minute or two an hour, all at the risk of losing time by getting pulled over or getting into an accident. The point is simple: ***Speeding is not worth it!***

Avoid the temptation to speed by planning ahead. Make the best use of your time so you won't feel any need to speed. Plan your trips with care to avoid the need to make up time. Get started on time. Maintain a steady pace. Plan rest and meal stops. Drive the speed limit.

experience speaking

"It's bad for the public image. When you're in a small car and see this big truck behind you, even though it's way behind you, that small car is... That truck's bigger, let's face it. It scares people. They slam on the brakes, and that truck can't stop as fast as a car can. You need to allow ample room and not intimidate people on the highways by being so close behind them."

Roberta
Safety Director
Florida

managing speed review

1. What three things add up to total stopping distance?

2. If you go twice as fast, will your stopping distance increase by twice or by four times?

3. Empty trucks have the best braking: True or False?

4. What is hydroplaning?

5. What is black ice?

6. What is meant by perception distance?

7. What is meant by reaction distance?

Managing Space 6 – 4

space ahead

space behind

space
to the sides

space overhead

space below

turning space

crossing &
entering traffic

Managing Space 6 – 4

To be a safe driver, you need a space cushion all around your vehicle.

Some drivers call it their "living room." When things go wrong, space gives you time to think and to take action.

To have space available when something goes wrong, you need to manage your space. While this is true for all drivers, it's very important for drivers of large vehicles. They take up more space on the road, and they require more space for stopping and turning.

Space Ahead

Of all the space around your vehicle, it's the area ahead of the vehicle — the space you're driving into — that's most important. You need space ahead in case you must stop suddenly. According to accident reports, trucks most often run into the vehicle in front of them. The most frequent cause is following too closely. Remember, if the vehicle ahead of you is smaller than yours, it can probably stop faster than you can. You may crash if you're following too closely.

How much space should you keep in front of you? One good rule says you need at least one second for each 10 feet of vehicle length at speeds below 40 mph. At greater speeds, you have to add one second for safety. For example, if you're driving a 40-foot vehicle, you should leave four seconds between you and the vehicle ahead.

In a 60-foot rig, you'll need six seconds. Over 40 mph, you'd need five seconds for a 40-foot vehicle and seven seconds for a 60-foot vehicle.

To know how much space you have, wait until the vehicle ahead passes a shadow on the road, a pavement marking or some other clear landmark. Then count off the seconds like this: "one thousand-and-one, one thousand-and-two" and so on, until you reach the same spot. Compare your count with the rule of one second for every 10 feet of length. If you are driving a 40-foot truck and only counted up to two seconds, you're too close. Drop back a little and count again until you have four seconds

experience speaking

"When you are trying to merge coming out onto the entrance ramp, you want to remember that you want to stay back from the vehicle in front of you. Don't get too close, because if you get too close, and they decide to stop, then you've got a problem."

Dave
Instructor
Wisconsin

of following distance (or five seconds, if you're going over 40 mph). After a little practice, you'll know how far back you should be. Remember to add one second for speeds above 40 mph. Also remember that when the road is slippery, you need much more space to stop.

Space Behind

You can't stop others from following you too close. But there are things you can do to make it safer. Stay to the right. Deal with tailgaters safely, and, of course, don't be one yourself.

Stay to the right. Heavy vehicles are often tailgated when they can't keep up with the speed of traffic. This often happens when you're going uphill. If a heavy load is slowing you down, stay in the right lane if you can. Going uphill, you should not pass another slow vehicle unless you can get around quickly and safely. "Quickly and safely" most often will mean when you have a passing lane to use.

In a large vehicle, it's often hard to see whether a vehicle is close behind you. You may be tailgated when you're traveling slowly. Drivers trapped behind slow vehicles often follow closely. In bad weather, many car drivers follow large vehicles closely, especially when it's hard to see the road ahead. There is an often misplaced sense of security for the driver of a smaller vehicle in sticking with the big rigs when the weather gets bad.

If you find yourself being tailgated, here are some things you can do. Avoid quick changes. If you have to slow down or turn, signal early and reduce your speed gradually. Increase your following distance. Opening up room in front of you will help you to avoid having to make sudden speed or direction changes. It also makes it easier for the tailgater to get around you. Don't speed up. It's safer to be tailgated at a low speed than a high speed. Avoid tricks. Don't turn on your tail lights or flash your brake lights.

Space To The Sides

Commercial vehicles are often wide and take up most of a lane. Safe drivers manage what little space they have. You can do this by keeping your vehicle centered in your lane, and avoid driving alongside others.

You need to keep your vehicle centered in the lane to keep safe clearance on either side. If your vehicle is wide, you have little room to spare.

There are two dangers in traveling alongside other vehicles: Another driver may change lanes suddenly and turn into you; You may be trapped when you need to change lanes.

Find an open spot where you aren't near other traffic. When traffic is heavy, it may be hard to find an open spot. If you must travel near other vehicles, try to keep as much space as possible between you and them. Also, drop back or pull forward so that you're sure the other driver can see you.

Strong winds make it difficult to stay in your lane. The problem is usually worse for lighter or empty vehicles. This problem can be especially bad coming out of tunnels. Don't drive alongside others if you can avoid it.

Space Overhead

Hitting overhead objects is a danger. Make sure you always have overhead clearance. Don't assume that the heights posted on bridges and overpasses are correct. Repaving or packed snow

may reduce clearances since the heights were posted. The weight of a cargo van changes its height. An empty van is higher than a loaded one. That you got under a bridge when your vehicle was loaded doesn't mean that you can do it when it's empty.

If you doubt you have safe space to pass under an object, go slowly. If you aren't sure you can make it, take another route. Warnings are often posted on low bridges or underpasses, but sometimes they are not.

Some roads can cause a vehicle to tilt because of the way they're constructed. There can be a problem clearing objects along the edge of the road, such as signs or trees. Where this is a problem, drive a little closer to the center of the road.

Before you back into an area, get out and check for overhanging objects, such as trees, branches or electric wires. It's easy to miss seeing them while you're backing. (Also check for other hazards at the same time.)

Space Below

Many drivers forget about the space under their vehicles. That space can be very small when a vehicle is heavily loaded. Railroad tracks, for instance, can stick up several inches. This is often a problem on dirt roads and in unpaved yards where the surface around the tracks can wear away. Don't take a chance on getting hung up halfway across. Drainage channels across roads can cause the end of some vehicles to drag. Cross such depressions carefully.

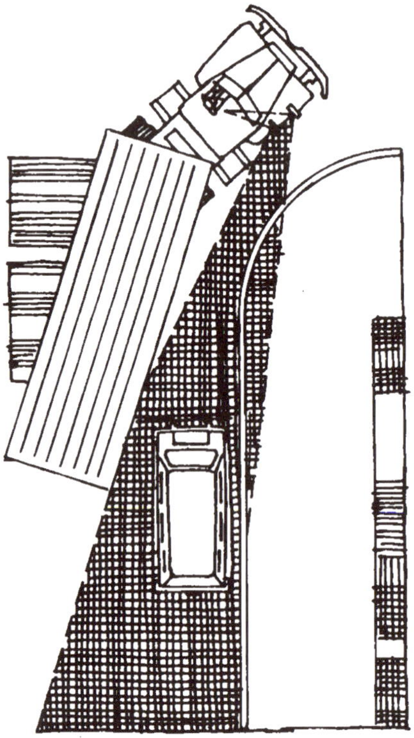

Turning Space

When turning, the tractor-trailer driver needs to check and double-check his space. Because of wide turning and offtracking, large vehicles are in danger of hitting other vehicles or objects during turns. Right and left turns each present their own set of problems.

right turns

Here are some rules to help prevent right-turn crashes. Turn slowly to give yourself and others more time to avoid problems. If you're driving a vehicle that can't make the right turn without swinging into another lane, turn wide as you complete the turn. Keep the rear of your vehicle close to the curb. While you can position your truck to make it hard for drivers to pass you on the right, never assume you have blocked traffic completely. Use your mirrors and keep checking all lanes.

Don't turn wide to the left as you start the turn. A following driver may think you're turning left and try to pass you on the right. You may crash into the other vehicle as you complete your turn. If you must cross into the oncoming lane to make a turn, watch out for vehicles coming toward you. Give them room to go by or to stop. However, never back up for them, because you might hit someone behind you. Again, use of mirrors is critical.

left turns

On a left turn, make sure you've reached the center of the intersection before you start the left turn. If you turn too soon, the left side of your vehicle may hit another vehicle because of offtracking.

If there are two turning lanes, always take the right hand turn lane. Don't start in the inside lane because you may have to swing right to make the turn. Drivers on your right may be hard for you to see. You may crash into them.

Crossing & Entering Traffic

Be aware of the size and weight of your vehicle when you cross or enter traffic. Because of slow acceleration and the space large vehicles require, you may need a much larger gap to enter traffic than you would in a car. Acceleration varies with the load. Allow more room if your vehicle is heavily loaded. Before you start across a road, make sure you can get all the way across before traffic reaches you. Timing to take advantage of the gap is very important.

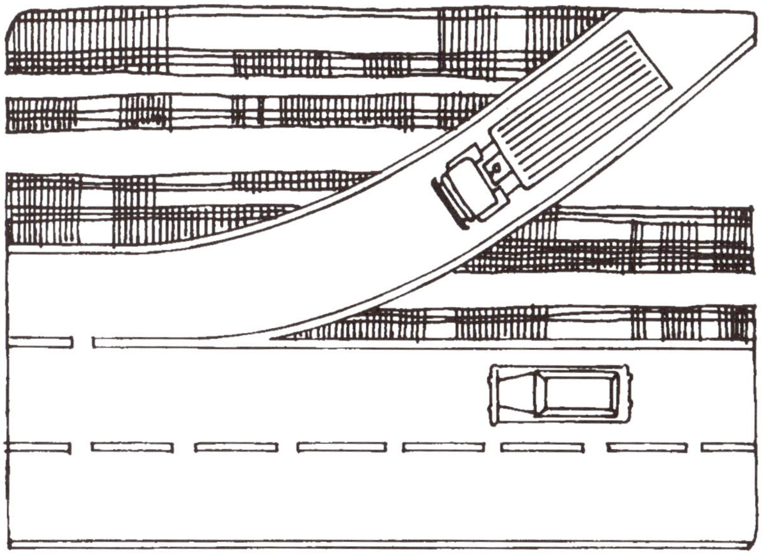

managing space review

1. How do you find out how many seconds of following-distance space you have?

2. If you are driving a 30-foot vehicle at 55 mph how many seconds of following distance should you allow?

3. You should decrease your following distance if somebody is following you closely: True or False?

4. If you swing wide to the left before turning right, another driver may try to pass you on the right: True or False?

5. What is being described when a driver refers to his “living room”?

6. When is it a good idea to pass another driver on a hill?

7. What are three common overhead dangers?

Driving at Night 6 – 5

driver factors

roadway factors

vehicle factors

night driving procedures

night driving adjustments

Driving at Night 6 – 5

Driving at night is more dangerous than driving during the day.

More than one-half of all traffic accidents occur at night. And, most fatal accidents occur between sunset and sunrise. Overall, there is less traffic travelling fewer miles at night. But a significant number of tractor-trailer miles are put on at night.

The dangers are obvious:

Low illumination makes things harder to see;

Hazards are much nearer before they can be seen;

A more limited sight area leaves less time to respond;

Drivers caught by surprise are less able to avoid hazards.

Problems caused by night driving involve all three elements of highway traffic: driver, roadway and vehicle. Each needs to be discussed separately.

Driver Factors

vision

People can't see as sharply at night or in dim light. Also, the eyes need time to adjust to seeing in dim light. You've probably noticed this when walking from the lobby into a dark movie theater or in other similar situations.

Furthermore, night vision is more affected by age than daylight vision. As we get older, the more light we need to see at night.

Take care of your eyes. Don't expose them to lights that are too bright. Wearing sunglasses on bright days can benefit your night vision by subjecting your eyes to less strain.

glare

Drivers can be temporarily blinded for a short time by bright light. It takes time to recover from this blindness. Older drivers are especially bothered by glare. Most people have been temporarily blinded by camera flash or by the high beams of an oncoming vehicle.

Even two seconds of glare blindness can be dangerous. A vehicle going 55 mph will travel more than half the distance of a football field during that time. Don't look directly at bright lights when driving. Look at the right side of the road. Watch the sidelines when someone coming toward you has very bright lights.

fatigue

Fatigue and lack of alertness are bigger problems at night. The body's need for sleep is beyond a person's control. Most people are less alert at night, especially after midnight. This is particularly true if you have been driving for a long time. Drivers may not see hazards as soon or react as quickly, so the chance of a crash is greater. If you're sleepy, the only cure is to get off the road and get some sleep. If you don't, you risk your life and the lives of others.

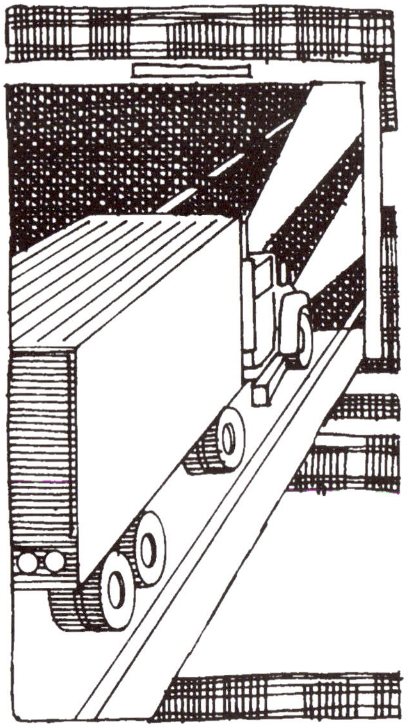

Roadway Factors

poor lighting

In the daytime, there's usually enough light to see well. This isn't true at night. Some areas may have bright street lights, but many areas will have poor lighting. On most roads, you'll probably have to depend entirely on your headlights.

Less light means you won't be able to see hazards as well as in daytime. Road users without lights are hard to see. Accidents at night involving pedestrians, joggers, bicyclists and animals are all too common.

Even when there are lights, the road scene can be confusing. Traffic signals and hazards can be hard to see against a background of signs, shop windows and other lights.

Drive slower when lighting is poor or confusing. Drive slowly enough to be sure you can stop in the distance you can see ahead.

drunk drivers

Drunk drivers and drivers under the influence of drugs are a hazard to themselves and to you. Be especially alert around closing times for bars and taverns. Watch for drivers who have trouble staying in their lane or maintaining speed, stop without reason, or show other signs of being under the influence of alcohol or drugs.

Vehicle Factors

headlights

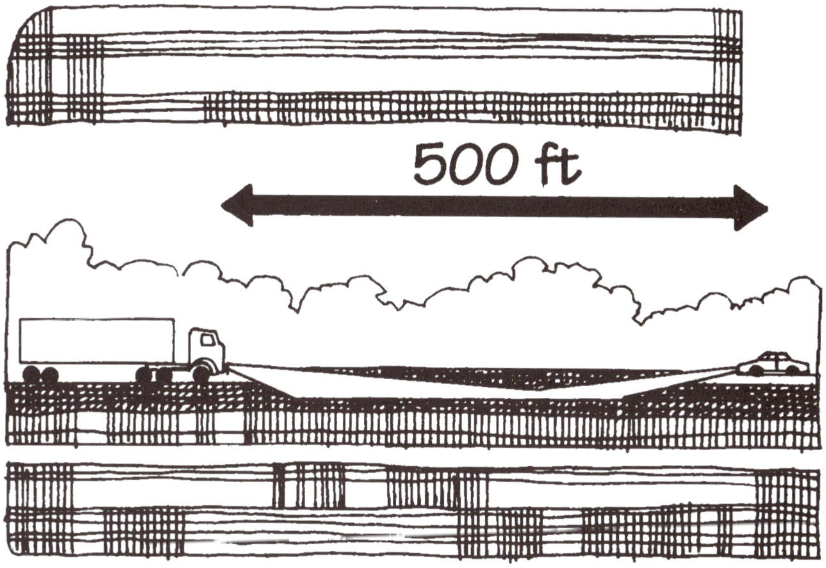

At night, your headlights will usually be the main source of light for you to see and for others to see you. You can't see nearly as much with your headlights, however, as you can see in the daytime. With low beams, you can see ahead about 250 feet and with high beams about 350-500 feet in good weather.

You must adjust your speed to keep your stopping distance within your sight distance. This means going slow enough to be able to stop within the range of your headlights. Otherwise, by the time you see a hazard, you won't have time to stop.

Night driving can be more dangerous if you have problems with your headlights. Dirty headlights may give only half the light they should. This cuts down your ability to see, and makes it harder for others to see you. Make sure your lights are clean and working.

Headlights can be out of adjustment. If they don't point in the right direction, they don't give you a good view and they can blind other drivers. Have a qualified person make sure they're adjusted properly.

experience speaking

"Night driving: Granted, you're going to get fatigued, at least from my own experience, as it gets to be later and later at night. You get even more fatigued as you get into the morning hours. You've still got to remain very aware of animals that may run out in front of you, possibly vehicles sitting along side the roadway with no flares, no triangles, no warning at all. And all of a sudden out of the night here they loom. You have to be constantly aware that there are obstacles out there just beyond your headlights. If you're fatigued, or tired, it doesn't pay to push it. You're just asking for it."

John
Driver
Wisconsin

other lights

In order for you to be seen easily, all lights or lighted areas must be clean and working properly: Marker lights, reflectors, clearance lights, taillights, identification lights.

At night, your turn signals and brake lights are even more important for communicating your intentions to others. Never drive with dirty or poorly functioning lights. Driving in snow means you will have to stop frequently to clear taillights.

windshields and mirrors

It's more important to have clean windshields and mirrors at night than in the daytime. Bright lights at night can cause dirt on your windshield or mirrors to create a glare of its own, blocking your view. Most people have experienced driving toward the sun just as it has risen or is about to set and found that they can barely see through a windshield that seemed to look okay in the middle of the day. Clean your windshield on the inside and outside regularly for safe driving at night.

Night Driving Procedures

Make sure you're rested and alert. Federal regulations are based on the idea that only rested drivers are safe drivers. If you're drowsy, sleep before you drive! Even a nap can save your life or the lives of others. If you wear eyeglasses, make sure they're clean and unscratched. Don't wear sun glasses at night. Do a complete pre-trip inspection of your vehicle. Pay attention to checking all lights and reflectors, and clean those you can reach.

Avoid blinding others. Glare from your headlights can cause problems for drivers coming toward you. They can also bother drivers going in the same direction you are, when your lights shine in their rearview mirrors. Dim your lights before they cause glare for other drivers. Dim your lights within at least 500 feet of an oncoming or leading vehicle.

Don't look directly at the lights of oncoming vehicles. Look slightly to the right. If other drivers don't put their low beams on, don't try to "get back" at them by putting your own high beams on. This increases glare for oncoming drivers and increases the chance of a crash.

Use your high beams when you can. Some drivers make the

mistake of always using low beams. This seriously cuts down on their ability to see ahead. Use high beams when it's safe and legal to do so. Use them when you're not within 500 feet of an approaching vehicle.

Also, don't let the inside of your cab get too bright. This makes it harder to see outside. Keep the interior light off. Adjust your instrument lights as low as you can and still read the gauges.

If you get sleepy, stop driving at the nearest safe place. People often don't realize how close they are to falling asleep even when their eyelids are falling shut. If you can safely do so, look at yourself in a mirror. If you look sleepy, or you just feel sleepy, stop driving! You are in a very dangerous condition. Sleep is the only cure for someone who's tired.

Develop the visual scan. If you have trouble scanning, you are probably too tired to keep driving.

experience speaking

"Don't overdrive your headlights. If you're in a rural area and your headlights shine a certain distance, count in seconds, one thousand and one, one thousand and two, and if you cross that point, then you may determine, hey, I may be going a little too fast, or I can pick up on my speed five to ten miles an hour. But don't overdrive your headlights. Do not look into a set of lights coming at you. It takes a typical person anywhere from two to seven seconds for their eyes to recover."

Gerry
Driver
Illinois

Night Driving Adjustments

For the most part, communicating at night is limited to your vehicle's lights and horns. Signalling your intentions is more critical at night. Signal all stops, slowdowns and direction changes a little earlier than you would during daylight. Use the horn lightly as necessary. Avoid blinding others with lights as a way of signalling.

Increase following distances at night by at least one second. Everyone needs more time and space to react.

Keep your speed within your sight distance. Don't overdrive your headlights. If you overdrive your headlights, you will end up going off the road when a curve comes up. It only takes a second to lose control of your vehicle if you try to adjust to an error too quickly.

Failing to adjust your speed at night will result in a loss of the time and space it takes to react to hazards. Many drivers have learned this the hard way on the road.

driving at night review

1. You should use your high beams whenever you can: True or False?

2. What should you do before you drive if you're drowsy?

3. Identify four common dangers in driving at night.

4. Why are drunk drivers a greater hazard at night than during daylight hours?

5. It's a good idea to keep your cab brightly lit at night: True or False?

6. What is meant by overdriving your headlights?

Extreme Driving Conditions 6 – 6

driving in winter

rain & fog

hot weather driving

uphill/downhill driving

night driving adjustments

Extreme Driving Conditions 6 – 6

The subject of extreme driving conditions is broad, covering a range of varying topics.

Generally extreme driving conditions can be divided into three categories: Weather; Terrain; and Traffic. While traffic creates the extreme driving conditions you'll most often encounter, the material that follows will deal with the situations brought about by weather and terrain. Traffic is dealt with elsewhere in this manual.

Further defining our subject areas, extreme weather conditions refers to winter driving, hot weather driving, rain and fog. When we discuss terrain, we are talking mainly about mountain or up-hill/down-hill driving and the affect of gravity on your vehicle.

No matter what the extreme driving condition, the truck driver will find it a challenge to himself and his vehicle. Both must be in top form.

Driving In Winter

Driver awareness and knowledge of vehicle components, systems and operating procedures are important in normal, everyday driving. They're critical when operating in the winter. Reduced traction increases stopping distances and decreases maneuverability and control. Reduced visibility from blowing snow means you won't see hazards as soon as you otherwise might, therefore giving you less time to respond.

You must make sure you and your vehicle are well prepared when you venture onto the road in winter. In addition to the regular pre-trip and on-the-road inspections you'll perform, you'll need to pay particularly close attention to the following items.

coolant level & antifreeze concentration

Make sure the cooling system is full and there's enough antifreeze in the system to protect against freezing. This can be checked with a coolant tester designed for that purpose.

defrosting & heating equipment

Make sure the defrosters work. They are an absolute necessity for safe driving. Make sure the heater is working and that you know how to operate it. If you use other types of heaters and expect to need them (e.g., mirror heaters, battery box heaters, fuel tank heaters), their operation must be checked, too.

wipers & washers

Make sure the windshield wiper blades are in good condition. Wiper blades have to press against the window hard enough to wipe the windshield clean. Otherwise, they may not sweep snow off properly. Make sure the windshield washer works and there's washer fluid in the washer reservoir. Use windshield washer antifreeze to prevent freezing of the liquid. If you can't see well enough while driving (for example, if your wipers fail), stop safely and fix the problem before you continue.

tires

Check tire tread depth. The drive tires must provide enough traction to push the rig over wet pavement and through snow. The steering tires must have traction to steer the vehicle. Tread depth is especially important in winter conditions. You should have at least 4/32 inch in every major groove on front wheels and at least 2/32 inch on other wheels. More would be better. Use a gauge to determine if you have enough tread for safe driving; don't guess.

tire chains

You may find yourself in conditions where you can't drive without chains, even to get to a place of safety. Prepare for the worst. Carry the right number of chains and extra cross links. Make sure they'll fit your drive tires. Check the chains for broken hooks, worn or broken cross links, and bent or broken side chains. Also check to see that chain slack adjusters are available if you'll need them. Learn how to put the chains on before you need to do it in snow and ice.

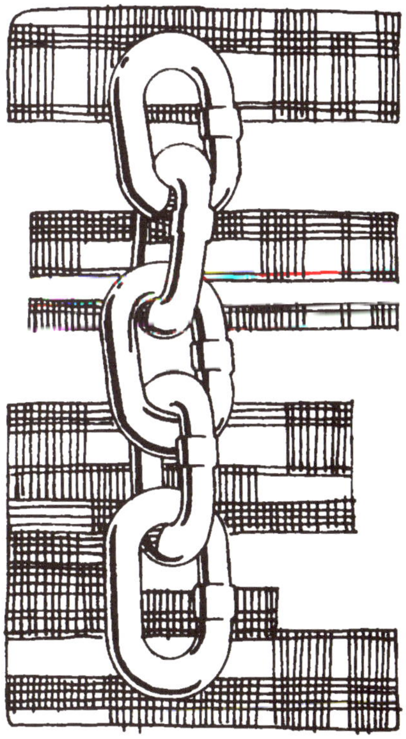

brakes

Have the brake balance checked. Brakes should apply in sequence from the rear forward. Check adjustment, and take up any slack. This limits air pressure to the brake chambers and helps prevent front wheel lockup. In winter driving conditions, you will need to make regular checks for ice on the brake linings, which can reduce braking power or even cause shoes to freeze to the drums when you're stopped.

Another important item is to keep air tanks as moisture-free as possible. Drain both tractor and trailer tanks daily. Moisture in the air lines will cause your brakes to freeze. If you're driving a vehicle with moisture-control equipment (e.g., alcohol evaporates, spitter valves), they have to be working properly. Check them out before you need them.

lights & reflectors

Make sure the lights and reflectors are clean. Visibility of lights and reflectors is especially important during bad weather. Check from time to time during bad weather to see that they're clean and working right.

windows & mirrors

Remove any ice, snow and debris from the windshield, windows and mirrors before starting. Use a windshield scraper, snow brush and windshield defroster as necessary.

hand holds, steps & deck plates

Remove all ice and snow from hand holds, steps, and deck plates used to enter the cab or to move about the vehicle. This will reduce the danger of slipping and falling.

winterfront

Make sure the winterfront isn't closed too tightly. If the shutters freeze shut or the winterfront is closed too tightly, the engine may overheat and stop.

exposed wiring & air lines

Ice and snow must be removed from wiring and air lines. Since snow and ice build-up can cause them to sag, even snag in tire chains, make sure they're properly supported.

fuel tank

Your fuel tank should be topped off before you start out and kept topped off regularly if bad weather is expected. This will help to keep water out of your fuel, which can freeze in fuel lines and filters. Filling the tank at the end of a trip and draining water from the bottom of fuel tanks are necessary precautionary measures in winter.

exhaust system

Exhaust system leaks are especially dangerous when cab ventilation may be poor (windows rolled up, etc.). Loose connections could permit poisonous carbon monoxide to leak into your vehicle. Carbon monoxide gas will make you sleepy. In large enough amounts, it can kill you. Check the exhaust system for loose parts and for sounds and signs of leaks.

coupling devices

You should double check the locking mechanism. Make sure the fifth wheel is lubricated with winter-grade lubricant. This will prevent binding and help steering on slippery roads.

interaxle differential lock

If your vehicle is equipped with an interaxle differential lock, you will need to check for proper orientation. If it's air operated, check for moisture in the air lines. And, of course, be aware of your operator manual's instructions for proper operation.

emergency gear

Always check for basic emergency gear before starting out when adverse weather conditions are present or expected. A scraper and snow brush are essential for keeping the windshield, windows, lights and reflectors clear of snow and ice. A small shovel should be on-board in the event that you become stuck or need to come to the aid of a stranded motorist. A military surplus type of shovel works well for most purposes.

Don't ignore the need for personal protective gear. Should you be stranded, you'll need drinking water, extra high-energy foods and any prescribed medicines you require. Also carry an extra pair of pants, boots and gloves. A hat, heavy jacket and coveralls will come in handy as well. The extra space these items take up initially will be a minor consideration in emergency situations. Driving in winter without proper equipment and clothing can be deadly.

experience speaking

"I think the biggest thing is knowing your equipment, knowing what your capabilities are. This is where most drivers are lacking. They don't realize what they're capable of doing or what the unit will do if it's handled properly. Given the right input, you can do a lot with that vehicle."

Jean
Driver
Iowa

weather reports

A good driver keeps himself aware of the latest weather and road conditions. If possible, listen to National Weather Radio Service broadcasts (162.4-162.55 MHz). This is a special radio band and is not found on your AM or FM radio. They're continually updated for numerous areas around the country. Plan your trip accordingly. Some areas, such as the eastern and western mountainous states, and the upper midwest can be particularly volatile, with conditions changing rapidly.

tire chain installation & use

Tire chains are constructed to grip the tire around the sides and provide traction across the tread. Some chains have V-lugs or cleats welded across them to deliver even greater traction. Single tire chains are attached to the outside tire on a set of duals. Doubles (or duals) cover a complete set of dual tires.

Chains can make the difference between going and not going. They're a must in many areas. Their ability to increase traction is dramatic; as much as 500%. Installed on the drive wheels, they improve uphill pulling power. On the trailer wheels, they improve downhill braking. It's important to mount chains at the first sign of slippery conditions in order to avoid stalling, skidding and, especially, jackknifing. Chains will help keep you on the road, and they'll decrease your chance of getting stuck.

Drivers should know the requirements -- regarding chains -- of the states they operate in. For instance, many mountainous states have strict and specific requirements for chain use under varying weather conditions. Location requirements may vary from district to district, as well as from state to state.

Chain-control areas are areas where, under certain road conditions, it's illegal to run without chains. Stations are usually set up ahead of chain-control areas to screen for the proper number of chains and placement. If chains aren't on-board in such areas, your vehicle will not be allowed to pass. Prepare your vehicle for the worst possible conditions, and be aware of the regulations.

Chains are most effective during periods of heavy snow and freezing rain. During a light, dry snow, they provide little traction advantage, but they do provide some stability. Most chains aren't effective on glare ice.

The thought of installing tire chains is an unwelcome one to many new drivers, but it need not be. Once mastered, the process is relatively easy. Chains should be snug, but not too tight. They are designed to creep or move on the tires to prevent gouging and breakage. After every five miles, chains should be

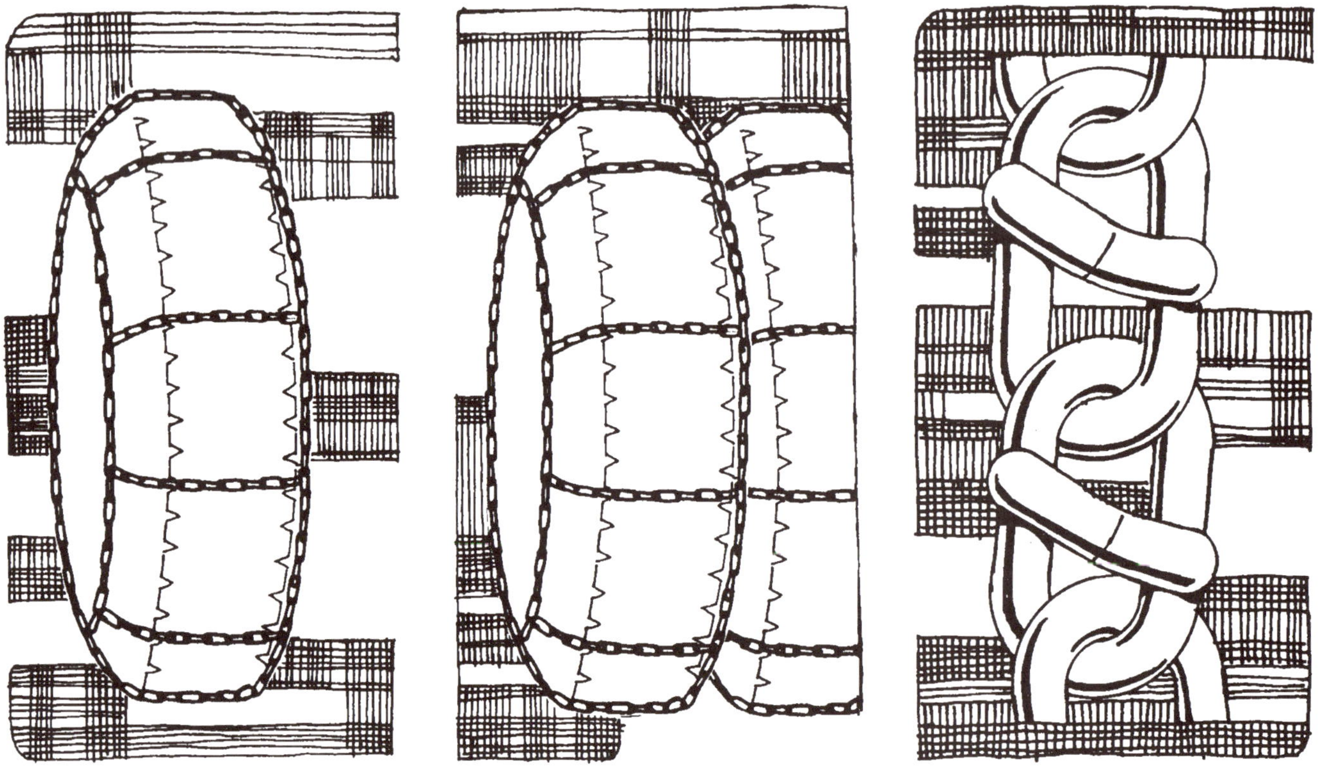

retightened to prevent slap against the trailer or catching on the suspension or fuel tank.

When you're installing chains, exercise extreme caution. Pull well off the road. Park on a level surface. Work facing traffic. Know where you're going to go if you're confronted with an out-of-control vehicle. Watch your footing. And, finally, watch your own rig: Vehicles with hot tires will sometimes slide on slippery surfaces.

A couple of additional precautions when running with chains do need to be noted. Again, exercise extreme caution. Vehicles running on snow are quiet, even with chains. New snowfall absorbs sound. As mentioned earlier, hot tires and chains melt snow, causing the surface to become very slippery. The result is that your vehicle may slide sideways while it's parked.

cold weather starting

A number of aids and procedures are available, ranging from additives to installed equipment, with a healthy dose of procedural know-how.

Ether and other ether-based fluids come in a variety of forms; capsules or "jelly beans," as they are commonly known; aerosol sprays; pressurized cylinders; and driver controlled injection systems. Each of these is meant, by design, to be administered either manually or automatically.

When using capsules, each provides one start. Capsules are placed in a special receptacle attached to the air cleaner. Don't put capsules in the air cleaner itself, as they can be sucked into the engine and cause severe damage.

Both aerosol sprays and pressurized cylinders require two persons for use. One spraying ether into a rag or other filtering device hung in front of the air cleaner and the second to start the vehicle. Not much of a spray is required. Over spraying may result in flashback or engine damage. The rag can also get sucked into the intake.

Vehicles equipped with automatic injection systems require only the driver for cold weather starting. A pump meters a predetermined amount of starting aid into the intake manifold. The pump is actuated by either a cab mounted switch or automatically when the tractor is started. Automatic systems are designed to eliminate excessive ether doses that can cause engine damage.

Warm air is needed for fuel to ignite. Glow plugs are electric heating elements that warm air coming into the engine from the air intake. With the variety of engines available, it's recommended that you consult your operator's manual for instructions.

Each driver must know how to operate the equipment on the vehicle he's driving. Operation may include hand priming the pump to spray fuel into the combustion chamber. This procedure increases combustion chamber temperatures. The compressed air, in turn, ignites the fuel. It usually takes roughly 60 seconds for glow plugs to heat the air sufficiently. A word of caution: Never use ether and glow plugs at the same time.

Preheaters keep the engine warm while your vehicle is parked for the night. Most over-the-road units with preheaters use an in-block type of preheater, while immersion heaters are usually used on off-highway construction and mining equipment. In-block heaters fit into the freeze plug holes in the lower water jacket. They operate by plugging them in to an appropriate electrical outlet. Coolant is then heated to approximately 160°, and the heat is circulated throughout the engine. This eliminates warm-up, allowing for the use of normal starting procedures.

In extremely cold areas, coolant heaters are frequently supplemented with battery box heaters, oil sump heaters and fuel heaters. Before using any of the above, check with your operator's manual and maintenance personnel for proper operation.

If your engine won't start with the use of starting aids, check the fuel and electrical systems. Check the exhaust stack while cranking the engine. If there is no vapor or smoke, the engine may not be getting fuel. Continued cranking will just run down the battery. Check the fuel tank and lines for ice blockage. Check the fuel tank vent.

Your battery must be in the best possible condition or it will not operate at full efficiency in cold weather. If you experience electrical problems, check the battery for terminal corrosion, loose connections, cracks in cables and moisture on the cables.

When you have an air starter, another problem that may cause starting difficulties is an exhausted air supply. In such a situation, you will need to resupply air from either an air compressor or another tractor. Check the operator's manual on procedures for air supply.

operating hazards

The two primary adverse weather operating hazards are reduced visibility and reduced traction.

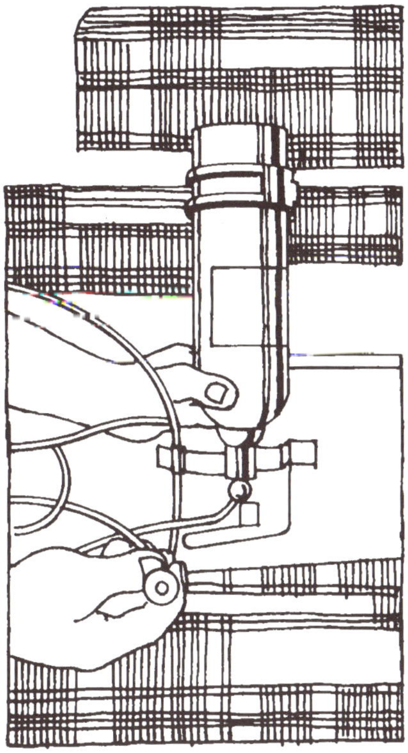

reduced visibility

In inclement weather, snow and ice tend to build up on windows and mirrors. When this happens, your visibility is reduced front, side and rear. Your defroster and windshield wipers, if operating properly, will keep the windshield clear in and outside. You must stop often, however, to clean off your rearview mirrors and side windows. Don't drive when you can't see in all directions.

Snow, ice and mud built up on lights and reflectors also reduce your visibility to other drivers. Clean them frequently. You need to see and be seen.

Fog, rain and even a medium snowfall can cut your visibility to 100 feet or less. A safe speed to allow for a 100 foot stopping distance is only 20 mph. Adjust your speed in bad weather to compensate for limited visibility.

Dense fog, a heavy downpour or blinding snow can cut visibility to near zero. When this happens, it's unsafe to drive at all. Pull off the road at the nearest safe place and stop until conditions improve.

reduced traction

The difference in road surfaces make for different degrees of friction, which in turn affects traction. For example: Dry concrete has greater traction than dry asphalt; wet concrete has greater traction than wet asphalt. Any snow packed or ice covered surface will have only 20% of the traction that same surface does when it's wet.

Drive wheels spin easily on slippery surfaces, impairing your ability to maneuver. This is especially true on cornering. Proper tire tread and weight on the drive wheels enhances traction and improves maneuverability. Making occasional light brake applications -- only when it's safe -- is one good way to check traction. Be alert to changing conditions.

Traction is needed for accelerating, turning and braking. As your speed increases, more traction is needed to perform any maneuver. If traction is poor, you must reduce your speed. On a wet surface, you may need to reduce your speed by one-fourth or more. As a general rule, you can drive about half your normal speed on packed snow. And on ice, you'll be lucky if you can travel at even one-third your normal speed.

Black ice is one of the most dangerous road conditions. It creates serious and sudden problems. In a sense, when there's black ice, the ice is camouflaged by the road. Black ice occurs when the ice is clear and can't be seen on the road surface. The most likely places you'll encounter black ice are on bridges, beneath underpasses, at dips in the road where water can collect and freeze, in shaded areas and on the lower side of banked curves. Black ice is particularly hazardous at night when visibility is further limited.

Finally, be alert for rain to turn to freezing rain when temperatures drop. Listen to a change in the sound of your tires and the road. Watch for spray from the tires of other vehicles: A lack of spray tells you that you're driving on ice. Feel for ice on the front of your outside mirror. If there's ice on it, there either is or soon will be ice on the road. If you choose to check traction by applying the brakes to see if your vehicle skids, do so gently and only when you are free of traffic on all sides.

skidding & jackknifing

The three basic causes of skidding and jackknifing are overacceleration, overbraking and oversteering. Too much acceleration will cause your drive wheel tires to spin. Overbraking will cause wheels to lock up. And oversteering may cause your trailer to skid or swing out. Don't try to accelerate out of a skid or jackknife. It will only make matters worse. Heavy braking will do no better. And hitting the brakes hard during a skid caused by oversteering is almost certain to result in a jackknife. A locked or spinning wheel wants to lead.

Skidding and jackknifing will be discussed further later in the section on Skid Control and Recovery.

slippery surfaces

It's important to drive slowly and smoothly on slippery roads. If it's very slippery, you shouldn't drive at all. Stop at the first safe place, and wait for conditions to improve.

When first starting, get the feel of the road. Don't hurry. Make turns as gentle as possible. Don't brake any harder than necessary, and don't use the engine brake or speed retarder. (They can cause the driving wheels to skid on slippery surfaces.)

Don't pass slower vehicles unless it's necessary. Go slow, and watch far enough ahead to keep a steady speed. Avoid having to slow down and speed up. Take curves at slower speeds and don't brake while in curves. Be aware that as the temperatures rise to the point where ice begins to melt, the road becomes even more slippery. Slow down more.

Don't drive alongside other vehicles. Keep a longer following distance. When you see a traffic jam ahead, slow down or stop to wait for it to clear. Try hard to anticipate stops easily and slow down gradually.

experience speaking

"When you sense that you're going into a skid, do not accelerate. Do not apply the brakes. The key element would be to steer, to counter steer, to try to keep it out. If you have a standard transmission, shove the clutch in, depress the clutch. A wheel rolling by itself tends to grab the pavement better than one that's spinning or sliding. Once you regain control of the vehicle, then you can do your braking, your stab braking, or whatever you have to do. But the key elements to regain control are to depress the clutch and counter steer."

Gary
Instructor
Wisconsin

wet brakes

While driving in heavy rain or deep standing water, your brakes will get wet.

Water in the brakes can cause the brakes to weaken, to apply unevenly or to grab. This can cause a lack of braking power, wheel lockups, pulling to one side or the other, and jackknife if you're pulling a trailer.

Avoid driving through deep puddles or flowing water if possible. If this can't be avoided, you should do the following. Slow down. Place your transmission in a low gear. Gently put on the brakes. This presses linings against brake drums or discs and keeps mud, silt, sand and water from getting in. Increase engine rpm, and cross the water while keeping light pressure on the brakes. When out of the water, maintain light pressure on the brakes for a short distance to heat them up and dry them out. Make a test stop when it's safe to do so. Check behind you to make sure no one is following, then apply the brakes to be sure they work right. If not, dry them out further as described above. Do not apply too much brake pressure and accelerator at the same time or you can overheat brake drums and linings.

experience speaking

"I've run pretty much exclusively west coast out of the Midwest myself and a lot across Interstate 80 and north of there in the winter. You encounter adverse driving conditions. Basically what I always try to do is discipline myself to think in slow motion. That every move I make I want to slow down. If I want to change lanes, I want to do it on a very gradual basis. If I have to stop, I want to allow myself that extra margin of room. It takes three to twelve times longer to stop on slick, wet, snowy pavement, icy pavement, than it does on a dry surface. You've got to allow yourself that time and discipline yourself to think in slow motion. I think it's helped me. I've got quite a few miles, and so far I haven't had an accident."

Dick
Driver
Minnesota

freeing a stuck vehicle

First and foremost, avoid conditions that might cause your vehicle to get stuck. Some obvious examples of such conditions would be soft road shoulders, deep snow, muddy roads and icy, slippery surfaces. Innocently pulling off the road to make repairs may find your vehicle stuck when the repairs are completed. Certain specific areas create more opportunity to get stuck. For instance, construction areas are generally difficult to get in and out of.

If your vehicle should become stuck, don't panic. Avoid spinning the drive wheels and rocking. This will only dig the vehicle in deeper. On ice, spinning wheels warm the ice under the tires and cut traction even further.

Use traction aids. Dig out from in front of the wheels. Scatter sand or gravel in the wheel path. Lay loose chains in front of the wheels. A number of traction enhancing products are available commercially.

When you're ready to try pulling out, start with your steering wheels facing straight ahead. Don't turn the wheels until your vehicle is moving. Starting out in second or third gear instead of first helps keep wheels from spinning and gives a smoother application of force.

Accelerate smoothly and gently. Sudden drive line shock may break axles, Universal joints or twist drive shafts. Ease off at the first sign of spinning or slipping. Allowing the wheels to continue spinning can dig you in deeper, cause spinout and damage your vehicle's differential.

It may take any or all of the above procedures to free your vehicle. When all else fails, you may need to call a tow truck.

towing

When towing is necessary, remember you're still responsible for your equipment and cargo. It's necessary that you, the driver, stay in charge, not the tow truck operator. It's your job to supervise the operation. If the tow truck operator does something wrong or unsafe, stop the operation, and correct the problem.

Hook the tow chain or cable to your vehicle yourself. Be sure enough chain or cable is played out so that your vehicle, once freed, will not lurch into the tow truck. In hooking up, pass the chain or cable through the hole in the bumper.

Do not hook it to the bumper.

Attach it to a solid portion of the frame or a frame cross member. (Some vehicles are equipped with tow hooks protruding from the frame.) Be very careful not to get the chain around the steering tie-rod or spring shackles along with the axle.

Before towing, agree on the procedure that will be followed. Know the direction the tow truck will pull and the direction your vehicle is to be steered. Agree on a signal that you can both clearly hear or see for stopping if you run into problems or for when your rig is clear.

The tractor should be accelerated gently; just enough to turn the wheels slowly. This will help you climb over obstructions. When you have been pulled clear, signal the tow truck driver to stop accelerating immediately. Apply tractor-trailer brakes, allowing your tractor-trailer to bring the tow truck to a stop. This will keep your rig from rear ending the tow truck.

A note of caution: Keep all bystanders well out of the way to prevent injuries should a chain or cable snap.

breakdown

Vehicle breakdowns are common in adverse weather. Be prepared. Exposure to wind and cold is dangerous and can easily result in frostbite. Stay in the cab.

Put on extra clothing to conserve body heat. Use your food and beverage supply cautiously in case help is slow in coming. It isn't unheard of for a vehicle to be marooned in remote areas for days before highway crews arrive. Don't try to walk for help unless it's absolutely necessary: You may not make it.

If you must leave your vehicle, leave a note on the steering wheel stating your destination and the direction you traveled.

Rain & Fog

The dangers of driving in rain or fog are often similar in nature to winter driving, and the precautions a driver must take are therefore similar to those covered earlier. There are, however, some situations unique to these conditions.

driving in rain

Highways are extremely slippery when first wet with rain. Be extra alert when driving at the beginning of a rainfall. Be prepared to reduce speed, open the space between your vehicle and others, allow more time to stop. Signal your intentions well in advance.

When rain begins to fall, it mixes with the oils that cover the road surface. Until this slippery mixture is broken down by more rain, the pavement is extremely slick. This condition can last a few minutes or a few hours. Even after the greasy film is gone, your vehicle's traction in rain is still not as good as on dry pavement.

When the road is wet, the least amount of moisture will be where vehicles ahead of you have thrown the water off the pavement and left tracks. Driving in these tracks will give you the best traction.

If rain is very heavy, your tires may have a tendency to ride-up on the standing water in front of them. This problem is called hydroplaning. The faster you drive, the more likely hydroplaning is to occur. Traction is only present when your tires are in contact with the road. If the tires are riding up on a wall of water, they lose traction. As traction decreases, the danger increases. If this happens, slow down.

driving in fog

When visibility is reduced, you must slow down. A safe speed in fog may be 20 mph if you can see six car lengths ahead, or 10 mph if you can see only two car lengths ahead. These are only general rules: The driver has to use his own best judgment as to what is safe.

Use your lights both to see better yourself and to help others see you. Your low beams work best in fog because water particles that make up fog reflect more light back at you when using high beams.

Low beams shine under the fog rather than into it and allow you to see the most. Yellow fog lights can help cut through fog, but extreme caution is still required.

Use your vehicle's windshield wipers. Part of the visibility problem in fog is caused by fine mist on the windshield. With your wipers on, the fog may not be as dense as it first appeared.

If you absolutely must stop, get completely off the road and turn on your flashers.

experience speaking

"Never worry about the other guys blowing by you, but drive what you feel safe at. If you feel safe driving at 30 miles an hour, slow down to 30 and then drop another five to ten miles off from that, and you'd be safe. I've found that worked many times for me. If I felt safe driving at 40 in a snow storm or in the fog, I'd slow down to 30. That gives you that extra little bit of room you need in case you do come upon the unexpected, someone parked in the fog, no lights on. You slow down that extra 5 to 10 miles per hour from what you feel safe at, and that gives you that extra cushion you need to be able to make that evasive maneuver if the unexpected comes up."

Gene
Instructor
Wisconsin

Hot Weather Driving

Problems encountered during winter driving or rain and fog are common, but it's important, too, to know something about the stresses of hot weather operation on both vehicle and driver as well. In extreme conditions, heat can be as hard on you and your equipment as cold. In addition to the regular pre-trip and on-the-road inspections you'll perform, you'll need to pay particularly close attention to the following items.

tires

Check tire mounting and air pressure. Inspect the tires every two hours or every 100 miles when driving in very hot weather. Air pressure increases with temperature. Don't let air out of the tires or the pressure will be too low when tires cool off. If a tire is too hot to touch, remain stopped until it cools off. Otherwise the tire may blow out or catch fire. Pay special attention to recapped or retreaded tires. Under high temperatures, the tread may separate from the body of the tire.

engine oil

Engine oil helps keep the engine cool, as well as lubricating it. Make sure there's enough engine oil. If you have an oil temperature gauge, make sure the temperature is within the proper range while you're driving.

engine coolant

Before starting out, make sure the engine cooling system has enough water and antifreeze according to the engine manufacturer's directions. (Antifreeze helps the engine under hot conditions as well as cold conditions.) While driving, check the water temperature or coolant temperature gauge from time to time. Make sure that it remains in the normal range. If the gauge goes above the highest safe temperature, there may be something wrong that could lead to engine failure and possibly start a fire. Stop driving as soon as safely possible and try to find out what's wrong.

Some vehicles have sight glasses or see-through coolant overflow containers or coolant recovery containers. These permit you to check the coolant level while the engine is hot. If the container isn't part of the pressurized system, the cap can be removed safely and coolant added even when the engine is at operating temperature.

Never remove the radiator cap or any part of the pressurized system until the system has cooled. On trucks, it's easy to be below the cap's height. Spraying of hot water and steam can cause severe burns to the face and eyes. If you can touch the radiator cap with your bare hand, it's probably cool enough to open. Always position yourself above the cap.

If coolant has to be added to a system without a recovery tank or overflow tank, follow these steps.

Shut the engine off. Wait until it's cooled. Protect your hands (use gloves or a thick cloth). Turn the radiator cap slowly to the first stop; this releases the pressure seal. Step back while pressure is released from the cooling system. When all pressure has been released, press down on the cap and turn it the rest of the way to remove it. Visually check coolant level, and add more coolant if necessary. Replace the cap, and turn it all the way to the closed position.

engine belts

Learn how to check V-belt tightness on your vehicle by pressing on the belts. Loose belts will not turn the water pump and/or fan properly. This will result in overheating. Also check belts for cracking, or other signs of wear.

hoses

Make sure coolant hoses are in good condition. A broken hose while driving can lead to engine failure and even fire. Look for breaks, cracks or wear.

driving

Watch for flash floods in desert areas. Secondary roads are frequently constructed through dry riverbeds. In heavy rain, the road can flood quickly. Try to avoid such areas if rain appears imminent.

Watch for bleeding tar. Tar or oils in pavement often rise to the surface in very hot weather. Such spots are very slippery.

Go slow enough to prevent overheating of tires and engine. High speeds create more heat for tires and engine. In desert conditions, heat may build up to the point where it's dangerous. The more heat, the greater chance of tire failure and fire, as well as an increased chance of engine failure caused by overheating.

breakdowns

Exposure to heat and sunlight is dangerous for the driver. Body fluids are used up rapidly in the desert, resulting in dehydration. Prolonged exposure can cause sunstroke. Stay out of the sun.

Don't stray from your vehicle if it breaks down. Stay in the cab or under the trailer. Wait for help to come.

Uphill/Downhill Driving

When we talk about uphill/downhill driving, we are usually talking about mountain driving, but any area where you encounter steep grades going up or down can be full of hazards and create extreme driving conditions.

In mountain driving, the force of gravity plays a major role. If you have a heavy load, you will have to use lower gears and go slower to climb hills. In coming down steep hills, gravity will tend to speed you up. You must go slow enough that your brakes can hold you back without getting too hot. If the brakes become too hot, they may start to fade. This means that you have to apply them harder and harder to get the same stopping power. This can start a reaction until you can't slow down or stop at all. These dangers can be avoided by going slow when going downhill.

Prior to starting mountain driving, drivers of vehicles equipped with air brake systems should check the following:

Compressor maintaining full reservoir pressure;

Pressure drop on full application within limitations;

Slack adjusters for push rod travel, within specifications;

Audible air leaks, applied and released;

Security of glad hands and air lines;

Drums for overheating (hold the back of your hand close to the drum, but do not touch it, or you could get burned);

Trailer protection valve operation.

Use of Gears

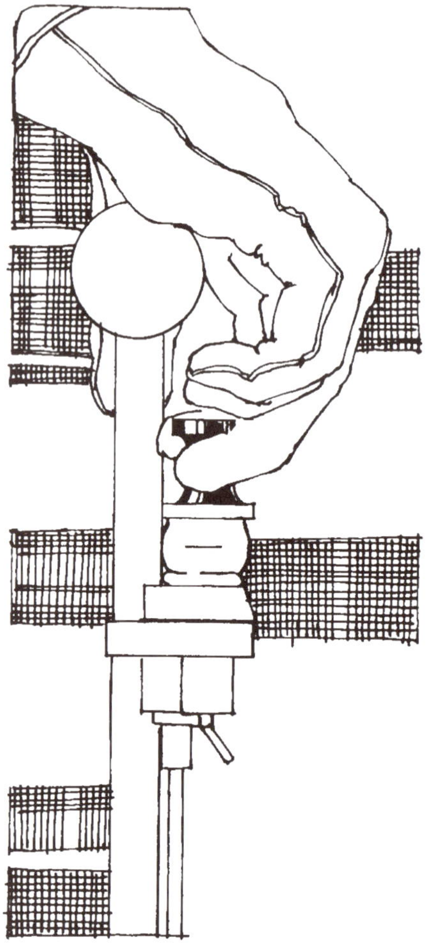

No matter what the size of your vehicle, going down long, steep grades can cause your brakes to fail if you go too fast. Using lower gears will help you keep from going too fast. Lower gears allow engine compression and friction to help slow the vehicle. This is true whether your transmission is manual or automatic.

If you do have a large vehicle with a manual transmission, don't wait until you've started down the hill to shift down. You might get hung up in neutral and lose the benefit of engine braking. You could find yourself coasting, which would be both illegal and dangerous. Be in the right gear before starting down the hill.

With older trucks, a rule for choosing gears was to use the same gear going down a hill that you would need to climb the hill. However, new trucks have low friction parts and streamlined shapes for fuel economy. They may also have more powerful engines. This means they can go up hills in higher gears and have less friction and air drag to hold them back going down hills. For that reason, drivers of modern trucks may have to use lower gears going down a hill than would be required to go up the hill. Find out what's right for your vehicle well in advance.

proper braking technique

When going downhill, brakes will always heat up. They're designed so brake shoes or pads rub against the brake drum or disks to slow the vehicle, which creates heat. Brakes are designed to take a lot of heat. However, brakes can fail from excessive heat by attempting to slow down from too high a speed too many times or too quickly. Brakes will fade (have less stopping power) when they get very hot, and can get to the point where they will no longer slow the vehicle.

An often repeated piece of advice for downhill braking has been to maintain a light and steady pressure on the brake pedal. Recent studies, however, suggest that intermittent and moderate downhill braking is safest.

The Americal Association of Motor Vehicle Administrators (AAMVA) recommends the following procedure.

1. Apply brakes just hard enough to feel a definite slowdown.
2. When speed is reduced to approximately five mph below the “safe” speed, release the brakes.
3. When speed picks up to the “safe” speed, repeat steps one and two as long as necessary to get down the grade.

auxiliary brakes/speed retarders

Auxiliary brakes, or speed retarders, reduce a vehicle’s forward motion without using the service brakes.

While relieving the burden on your vehicle’s service brakes, retarders work in conjunction with the service brakes to help you maintain control on long downgrades. There are several types of auxiliary brakes/retarders.

Engine brakes (such as the well-known “JakeBrake”) eliminate the power stroke and convert the engine to an air compressor. Fuel injection is stopped and exhaust valves open. Compressed air is expelled, piston movement is retarded.

Exhaust brakes back exhaust gasses up into the engine and create a 40 to 50 psi pressure, retarding piston movement.

Hydraulic retarders are either located between the engine and transmission or are an integral part of the transmission. They use engine or transmission oil to slow drive shaft or rotor rotation to slow the vehicle.

experience speaking

“I think the important thing to remember, there is an old rule of thumb, that you go down the hill in the same gear you went up the hill. If you have an engine in the 300 horsepower range, that probably is true, because you go up the hill rather slowly, probably 18 to 20 mile per hour with a full load. If you have a bigger engine , up in the 400 horsepower range, you’ve got to drop one gear more, because you can go up the hill faster than you should really come down the hill. If the truck is equipped with a jake brake or some type of retarder, you still have to act as if it doesn’t have it, as far as I’m concerned. Something could go wrong with it. There could be a wire shorted out. If you were coming down the hill at an excessive speed, and the jake stopped working, and you had to resort to the brakes, and you were halfway down, you might not be able to make it. I think caution is probably really the watchword. I’ve had a couple of smoking brakes on a truck, early on in my career, and I just really don’t want to repeat the process. You can feel the brakes starting to fade out as they get hotter and hotter until the point where there just isn’t anymore braking power left. I really don’t want to experience that.”

Dick
Driver
Wisconsin

Electric retarders are mounted in the drive line where electric current retards drive-shaft rotation.

Auxiliary brakes or speed retarders are desirable in any operation where service brakes are continually used, such as mountainous terrain or long descending grades. The driver must know when and when not to use retarders.

Noise ordinances prohibit their use in some locations. Proper gears, RPM and weather conditions are all important factors.

escape ramps

Escape ramps have been built on many steep mountain grades. They're intended to stop runaway vehicles safely without injury to driver or passengers.

There are four basic types of escape ramps. Gravity ramps are built on steep ascending grades with a pea gravel surface and mounds of gravel or sand at the end of the ramp. Sand piles have mounds or ridges high enough to drag against the undercarriage of vehicles. Arrester beds have masses of loose material, such as pea gravel, into which a vehicle sinks. Combination ramp and arrester beds, as their name applies, rely on a combination of loose surface material and long grades to slow vehicles.

Recognizing the safety role of escape ramps is vital. Drivers are confronted with two choices. You may use escape ramps, thus saving yourself (and possibly others), your vehicle and its cargo at the minimal cost of minor damage and a tow. Or, you may pay the consequences of not using ramps: Loss of cargo, loss of vehicle and worst of all, the loss of other lives in addition to your own.

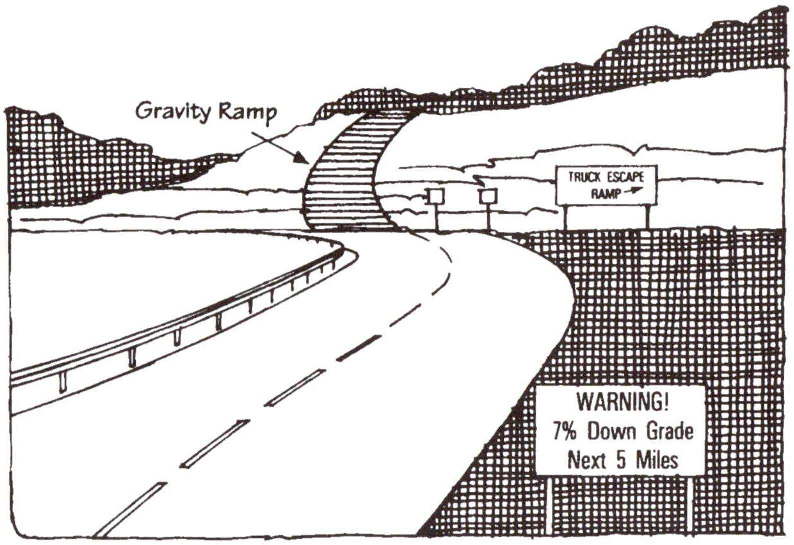

extreme driving conditions **review**

1. What effects can wet brakes cause, and how can you avoid these problems?

2. Should you let air out of hot tires so pressure goes back to normal?

3. You can safely remove the radiator cap as long as the engine isn't overheated: True or False?

4. Why should you be in the right gear before starting down a hill?

5. Why do newer trucks use lower gears going down a hill than coming up?

6. The key to preventing brake fade is to go slow enough for conditions: True or False?

7. Tire chains are sometimes a good idea, but they are never required by law: True or False?

8. What might you expect on a damp winter day when the temperature hovers around 32° F?

9. Who is in charge when your vehicle needs to be towed?

10. On a long downhill grade, is it better to pump your brakes lightly and often or maintain a light, steady pressure?

11. What is an escape ramp?

Seeing Hazards 6 – 7

road hazards

user hazards

have a plan

Seeing Hazards 6 – 7

A hazard is any road condition or other road user (driver, bicyclist, pedestrian) that's a possible danger.

For example, a car in front of you is headed towards the freeway exit, but his brake lights come on and he begins braking hard. This could mean that the driver is uncertain about taking the offramp. He might suddenly return to the highway. This car is a hazard. If the driver of the car cuts in front of you, it's no longer just a hazard; it's an emergency.

Seeing hazards lets you be prepared. It allows you to have a plan. You'll have more time to act if you see hazards before they become emergencies. In the example above, you might make a lane change or slow down to prevent a crash if the car suddenly cuts in front of you. Seeing this hazard gives you time to check your mirrors and signal a lane change. Being prepared reduces the danger. A driver who didn't see the hazard until the slow car pulled back on the highway in front of him would have to do something very suddenly. Sudden braking or a quick lane change is much more likely to lead to a crash.

There are often clues that will help you see hazards. The more you drive, the better you can get at seeing hazards.

Road Hazards

Road hazards appear in many different forms. They may be naturally occurring, man-made or a combination of the two.

Road hazards are the outstanding characteristics of the road surface, shape or contour that could adversely affect a driver's ability to control his vehicle or see clearly. Road hazards are a major factor in truck accidents.

work zones

When people are working on the road, it's a hazard. There may be narrow lanes, sharp turns or uneven surfaces. Other drivers are often distracted and drive unsafely. Workers and construction vehicles may get in the way. Drive slowly and carefully near work zones. Use your four-way flashers or brake lights to warn drivers behind you.

drop-off

Sometimes the pavement drops off sharply near the edge of the road. Driving too near the edge can suddenly tilt your vehicle toward the side of the road. This can cause the top of your vehicle to hit roadside objects (signs, tree limbs). Also, it can be hard to steer as you cross the drop-off, going off the road or coming back on.

foreign objects

Debris on the road can be hazards. They can be a danger to your tires and wheel rims. They can damage electrical and brake lines. They can be caught between dual tires and cause severe damage. Some obstacles which appear to be harmless can be very dangerous. For example, cardboard boxes may be empty, but they may also contain some solid or heavy material capable of causing damage. The same is true of paper and cloth sacks. It's important to remain alert for objects of all sorts, so you can see them early enough to avoid them without making sudden, unsafe moves.

offramps/onramps

Freeway and turnpike exits can be particularly dangerous for commercial vehicles. Offramps and onramps often have speed limit signs posted. Remember, these speeds may be safe for automobiles, but may not be safe for larger or heavily loaded vehicles. Exits which go downhill and turn at the same time can be especially dangerous. The downgrade makes it difficult to reduce speed. Braking and turning at the same time can be a dangerous practice. Make sure you're going slow enough before you get on the curved part of an offramp or onramp. A rule of thumb to follow is to go 5 mph slower than posted speed limit (and this is for ideal conditions).

User Hazards

In order to protect yourself and others, you must know when other road users may do something hazardous. Some clues to these types of hazards follow.

blocked vision

People who can't see others are a very dangerous hazard. Be alert for drivers whose vision is blocked. Vans, loaded station wagons and cars with the rear window blocked are examples. Rental trucks should be watched very carefully. Their drivers often aren't used to the limited vision they have to the sides and rear of the truck. In winter, vehicles with frosted, ice covered or snow covered windows are hazards.

Vehicles may be partly hidden by blind intersections or alleys. If you can only see the rear or front end of a vehicle but not the driver, then he can't see you. Be alert because he may back out or enter into your lane. Always be prepared to stop.

Delivery trucks can present a hazard. The driver's vision is

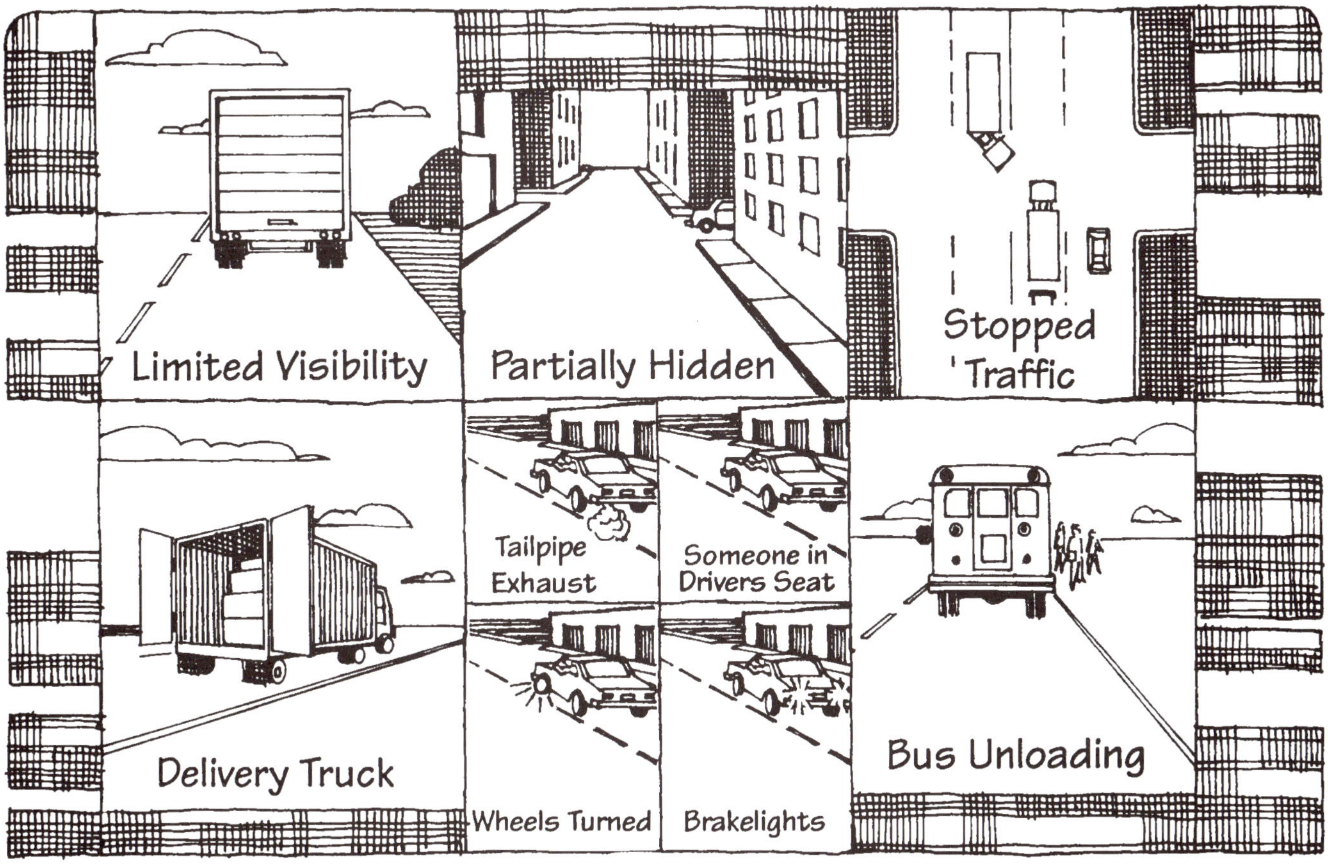

often blocked by packages or the vehicle doors. Drivers of step vans, postal vehicles and local delivery vehicles often are in a hurry and may suddenly step out of their vehicle or unexpectedly drive their vehicle into the traffic lane.

Parked vehicles can be hazards when people start to get out in traffic. Or they may suddenly start up and drive into your pathway. Watch for movement inside the vehicle or movement of the vehicle itself that shows people are inside. Watch for brake lights or backup lights, exhaust and other clues that a driver is about to move.

Be careful of a stopped bus. Passengers may cross in front of or behind the bus, and they often can't see you.

Pedestrians and bicyclists can also be hazards. Walkers, joggers and bicyclists may be on the road with their back to the traffic so they can't see you. Sometimes, they wear portable stereos with head sets so they can't hear you either. This can be dangerous. Bicyclists have the right to be on the road and you must yield to them. On rainy or snowy days, pedestrians may not see you because of hats or umbrellas. They may be hurrying to get out of the weather and may not pay attention to the traffic.

distractions

People who are distracted are hazards. Watch for where they're looking. If they're looking elsewhere, they can't see you. But be alert even when they are looking at you. They may believe that they have the right of way.

children

Children tend to act quickly without checking traffic. Children playing with one another may not look for traffic and are a serious hazard.

talkers

Drivers or pedestrians talking to one another may not be paying close attention to traffic. Keep an eye out for drivers who wave their arms or look more at their passengers than the road.

workers

People working on or near the roadway are a hazard clue. The work creates a distraction for other drivers, and the workers themselves may not see you.

disabled vehicle

Drivers changing a tire or fixing an engine often don't pay attention to the dangers of roadway traffic. They're often careless. Jacked up wheels or raised hoods are hazard clues.

accidents

Accidents are particularly hazardous. People involved in the accident may not look for traffic. Passing drivers tend to look at the accident. People often run across the road without looking. Vehicles may slow or stop suddenly.

shoppers

People in and around shopping areas are often not watching traffic, because they're looking for stores or looking into store windows. Shopping malls create particularly dangerous areas.

confused drivers

Confused drivers often change direction suddenly or stop without warning. Confusion is common near freeway or turnpike interchanges and major intersections. Tourists unfamiliar with an area can be very hazardous. Clues to tourists include cartop luggage and out-of-state license plates. Unexpected actions (stopping in the middle of a block, changing lanes for no apparent reason, backup lights suddenly going on) are clues to confusion. Hesitation is another clue, including driving very slowly, using brakes often, or stopping in the middle of an intersection. You may also see drivers who are looking at street signs, maps and house numbers. These drivers may not be paying attention to you.

experience speaking

"They have to exercise a lot of common sense. If you don't do that, if you make a bad decision, you're going to pay for it. There are other people on the roads as well. I wouldn't want to be the person that ran over a car and killed a family. I personally don't know how I'd live with myself if I did that. I encourage people to stop and just think about what they're doing in every move they make. They're driving a big vehicle. If they hit a car, a pedestrian, a motorcycle, in many cases, even another truck, the outcome is devastating to themselves as well as to the other people that are involved. It takes a certain amount of restraint."

Dick
Driver
Wisconsin

slow drivers

Motorists who fail to maintain normal speed are hazards. Seeing slow moving vehicles early can prevent a crash. Some vehicles, by their nature, are slow and seeing them is a hazard clue (mopeds, farm machinery, construction machinery, tractors, for example). Some of these will have the slow moving vehicle symbol to warn you. This is a red triangle with an orange center. Watch for it.

Drivers signaling a turn may be hazards. Drivers signaling a turn may slow more than expected or even stop. If they're making a tight turn into an alley or driveway, they may go very slow. If they're blocked by pedestrians or other vehicles, they may have to stop on the roadway. Vehicles turning left might have to stop for oncoming vehicles.

drivers in a hurry

Drivers — even drivers of other tractor-trailers — may feel your commercial vehicle is preventing them from getting where they want to go on time. Such drivers may pass you without a safe gap in the oncoming traffic, cutting too close in front of you. Drivers entering the road may pull in front of you in order to avoid being stuck behind you, causing you to brake. Be aware of this, and watch for drivers who are in a hurry.

impaired drivers

Drivers who are sleepy, had too much to drink, on drugs or ill are hazards. Impaired drivers may be weaving across the road or drifting from one side to another. They may slip off the shoulder or bump curbs on turns. If another driver stops at green lights or waits too long at a stop, keep an eye on them. Also be alert to drivers who continually speed up and slow down. Be on the look out for drunk or otherwise impaired drivers at all times.

driver body movement

Drivers look in the direction they are going to turn. You may sometimes get a clue from a driver's head and body movements that he's going to make a turn, even though the turn signals aren't on. Drivers making over-the-shoulder checks may be going to change lanes. These clues are most easily seen in motorcyclists and bicyclists. Watch other road users and try to tell whether they might do something hazardous. Try to make eye contact.

conflicts

You're in conflict when you have to change speed or direction to avoid hitting someone. Conflicts occur at intersections where vehicles meet, at merges (such as onramps) and where there are needed lane changes (such as the end of a lane, forcing vehicles to move to another lane of traffic). Other situations include slow moving or stalled traffic in a traffic lane and accident scenes. Watch for other drivers who are in conflict because they are a hazard to you. When they react to this conflict, they may do something that will put them in conflict with you.

Have A Plan

You should always be looking for hazards. Continue to learn to see hazards on the road. However, don't forget why you're looking for the hazards: they may turn into emergencies. You look for the hazards in order to have time to plan a way out of an emergency. When you see a hazard, think about the emergencies that could develop and figure out what you would do.

Keep asking yourself "what if" questions, such as "what if that car pulls out?" Always be prepared to take action based on your plans. In this way, you will be a prepared, defensive driver who will improve not only your own safety but the safety of all road users.

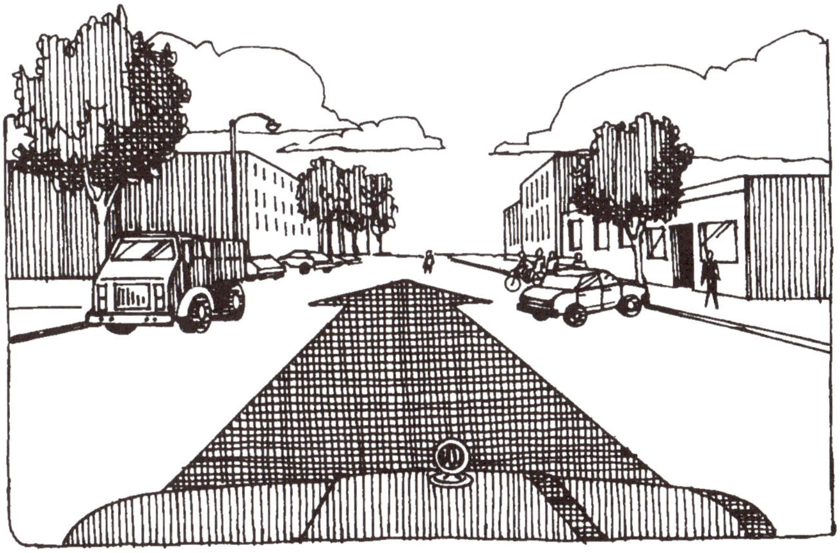

seeing hazards **review**

1. What is a hazard?

2. Why make emergency plans when you see a hazard?

3. What are some common road hazards?

4. What are some common user hazards?

5. Why are offramps and onramps particularly dangerous?

6. How can you spot a confused driver?

7. What are some clues that another driver is drunk or under the influence or drugs?

Emergency Maneuvers 6 – 8

evasive steering

emergency braking

offroad recovery

brake failures

blowouts

Emergency Maneuvers 6 – 8

The safest way to handle an emergency is to keep it from happening in the first place.

Most emergencies can be avoided by successful hazard perception and avoidance. The more emergencies a driver encounters, the greater risk of an accident. Given enough emergencies, sooner or later, the driver will have an accident.

Most emergencies are caused by driver error. Emergencies arise when one or more drivers fail to observe safe operating practices. Professional drivers reduce the likelihood of emergencies by employing all of their knowledge and skill, and by remembering that all road users are equal. It's fair to note that often common courtesy can prevent emergency situations. If you find you are having too many emergencies, evaluate your driving techniques.

Evasive Steering

A tractor-trailer combination can generally be turned more quickly than it can be stopped. If an escape path is available, evasive steering provides a better chance of avoiding a collision than attempting to stop. Head-on and rear-end collisions are often fatal. Evasive steering can often reduce damages to a side-swipe. Constantly scanning from mirror to mirror gives you the information to make a quick decision.

The two most common escape routes from an accident are another traffic lane and the road shoulder. If your lane is entered, a quick lane change would be the best escape route. If a lane change is impossible or dangerous, the shoulder of the road provides an alternative escape route. The ditch is the best alternative in some circumstances and should not be eliminated from your options.

Evasive steering, when handled properly, is generally safe. Quick, evasive maneuvers usually won't cause a roll-over for the practiced driver. They're not usually as dangerous as entering a curve or a turn too quickly. And even the small chance of a roll-over is better than a certain collision. Of course, the safest conditions for evasive steering are when you have stable cargo with a low center of gravity and a firm footing, such as an adjacent lane of highway or a paved shoulder.

general procedures

Minimize the amount of turning necessary. Start evasive steering as early as possible. Start as soon as an emergency is seen. The earlier an evasive maneuver starts, the smaller the amount of turn necessary.

Turn only as much as is needed. Aim to just clear the obstacle. The larger the turn, the greater the chance of a jackknife or rollover.

Turn as quickly as possible. Use hand-over-hand steering. Each turn of the wheel should be approximately 180 degrees. This is the best reason for driving with your hands in a nine o'clock-three o'clock position. This position allows you to turn the wheel 180 degrees without letting go. Your hands won't be in the right position in an emergency if they aren't always in the right position.

Avoid braking while turning. Braking could cause the tractor and trailer wheels to lockup. Locked wheels during turning could easily result in a loss of control.

Brake before turning. If distance permits, apply the brakes hard before beginning a turn. Braking as much as possible reduces your speed before turning. This allows your vehicle to be turned more sharply and reduces the chance of a jackknife or rollover.

You must be prepared to countersteer quickly. (Countersteering is the act of turning back toward your original path of travel.) Quick countersteering is required to keep your vehicle from going outside of its escape path and off the shoulder or into traffic.

Depending on the length of the trailer, begin to countersteer as soon as the front clears the obstacle.

At this point, the importance of wearing your seat belt is

magnified. It's difficult to turn the steering wheel quickly unless you're firmly rooted on the seat. A quick turn might otherwise cause you to slide, or even eject, out of your seat. You can't steer unless you're in the driver's seat. Loss of control becomes likely, a secondary collision almost certain. No seat belt means you are hanging on to the steering wheel and are not able to use the wheel.

The details of how an evasive maneuver is performed depend upon each specific situation. Certain procedures are needed for emergencies involving oncoming vehicles, stopped vehicles and merging vehicles.

oncoming vehicle

Evade an oncoming vehicle by steering to the right. Blast your horn to get the driver's attention. The oncoming driver will tend to turn to his right.

Don't steer to the left. Entering the left lane will probably cause a collision to occur in that lane. Leaving the road is preferable to a head-on collision.

stopped vehicle

If the left lane is clear, turn left into that lane. Swerving right creates the chance of sideswiping a vehicle alongside. The height of your cab should allow you to see if the left lane is clear. This is one situation in which it can be safe for you to turn into the left lane.

If you're next to a clear shoulder, you have the alternative of going right should the left lane be blocked.

If you're in the middle lane of a multilane road, your obvious choice must be to steer into whichever lane presents the least danger. Otherwise, evade to the right. It's better to force a vehicle there to the shoulder than to force another vehicle into an oncoming traffic lane.

converging vehicle

Blast the horn. This generally freezes the other driver and, hopefully, makes him stop quickly. Making the other driver stop usually minimizes the amount of evasive steering you'll have to do. Proper use of the horn might prevent many accidents that occur. The blast of your horn may frighten or annoy the other driver, but better that than to have a collision.

Swerve away from the converging vehicle. Don't try to steer behind it: If the other vehicle stops, a right angle collision will result. On the other hand, you shouldn't steer away from the converging vehicle if it puts your vehicle in the path of oncoming traffic. It's better to collide at an angle with a merging vehicle than to swerve into the path of an oncoming one.

Emergency Braking

technique

Over-application of the brakes will lockup the wheels, causing your vehicle to skid. A skid is likely to produce a jackknife. The tractor and trailer can actually collide in a jackknife. Either the tractor or trailer could collide with trees, buildings, other structures or, most especially, other vehicles. Loss of control during a jackknife can cause the entire rig to cross into other lanes.

Proper braking technique brings your vehicle to a stop in a minimum amount of distance. It also allows control to be maintained during a stopping maneuver.

procedure

Either of two emergency braking techniques can be used: Controlled braking or stab braking. Both should help prevent a skid or jackknife.

In controlled braking, you maintain a steady pressure on the brakes, applying them to just short of lockup. Developing skill in controlled braking technique requires practice in the vehicle you'll be operating. It's difficult to anticipate the point of lockup precisely. Without practice at controlled braking, stab braking is your best bet.

With stab braking on more or less level ground, you begin by applying the brake fully. Then release the brake partially when the wheels lock. Applying the brake achieves maximum braking, and releasing, or backing off the brakes, avoids a skid. When the wheels start to roll again, reapply the brake. Make sure you allow time for the wheels to roll again between each stab, because reapplying too quickly will cause a skid. Repeat the stab braking sequence until your vehicle has slowed enough for a safe stop or turn maneuver.

Offroad Recovery

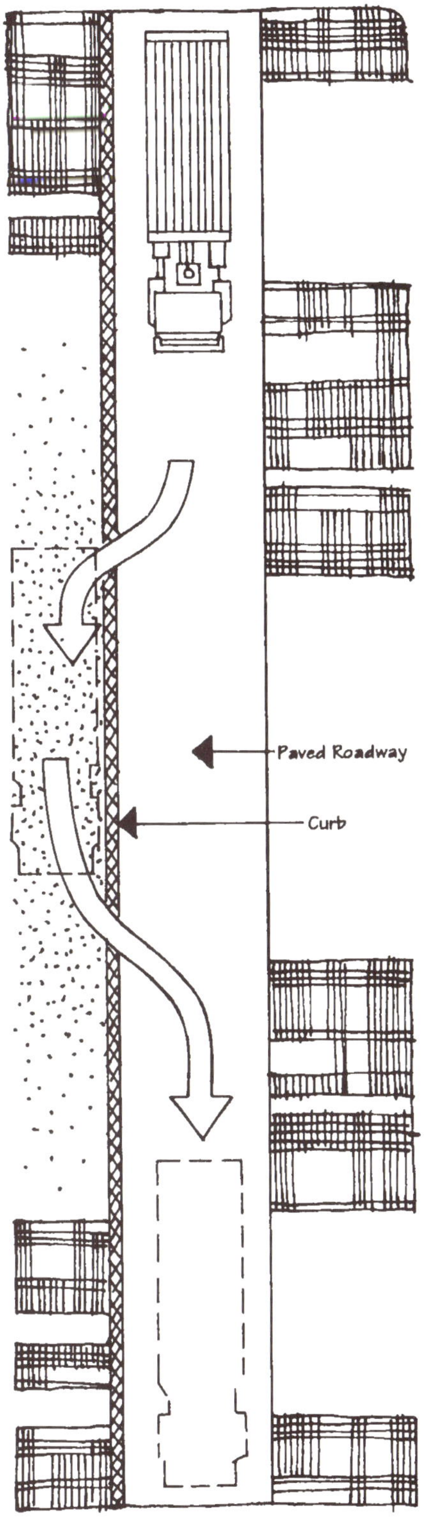

In many emergencies, the roadside will provide the best escape path. Whether on the right side of the road or on the median strip, where a shoulder is provided, use of the roadside is better than colliding with another vehicle.

Drivers are often fearful of leaving the roadway. Many have heard stories about crashes resulting from the use of the roadside for an evasive maneuver.

However, the fact is that many roadside crashes resulted from driver error; sometimes no other vehicle or object was even involved. Furthermore, successful evasive maneuvers don't result in an accident and so are not reported.

Successful offroad recovery often requires leaving the roadway immediately. Too often drivers fear leaving the roadway and react too late. Most accidents making use of offroad recovery are due to incorrect technique.

Offroad recovery is generally safe when the roadside is wide enough to accommodate a vehicle (for instance, no ditches or culverts) and firm enough to support a vehicle (soft mud, sand and gravel can cause problems).

procedure

Brake before turning. Reduce speed as much as possible. Use controlled or stab braking to prevent loss of control. Avoid braking while turning because the vehicle is more likely to skid during braking. And steering control is particularly important while entering the roadside, where gravel or dirt will reduce traction.

Minimize turning. Keep one set of wheels on the pavement if possible. This will aid traction and help you maintain control. Maintain as straight a course as possible. Each turn creates the danger of a skid.

If the roadside is clear, fight the tendency to return to the roadway. Grasp the wheel firmly and concentrate on steering.

Stay on the roadside until your vehicle comes to a complete stop. Allow engine compression to slow and help stop the vehicle. Apply your brakes only after speed has been reduced. When the situation has passed, signal and check your mirror before returning to the travel lane.

Sometimes the roadside will be blocked by a parked vehicle, telephone pole, sign, bridge abutment or other obstruction, and you'll need to return to the roadway more quickly. If this happens, allow your vehicle to slow as much as possible before returning to the road. Then turn the wheel sharply toward the roadway: Attempting a gradual return to the roadway is hazardous and can cause you to lose control suddenly. Turning sharply allows you to determine the point of return. You won't be caught by surprise, and you can be ready to countersteer.

Upon returning to the roadway, countersteer; turn quickly in the direction of the roadway. Turn as soon as the right wheel rides up on the surface. Both turning back onto the roadway and countersteering should be executed as a single, integrated steering maneuver.

Often, if a truck is driving too close to the edge of the roadway, the wheels will drop off the pavement. If this happens, avoid immediate return to the roadway. You can maintain control easily with two wheels on the pavement. A quick return has caused many vehicles to overturn or veer across the roadway.

When your wheels drop off the pavement, follow the same procedure as for offroad recovery.

Brake Failures

A well-maintained brake system rarely fails completely. Several fail-safe devices are designed to prevent brake failures from causing an accident. Despite this, however, brake failures do occur. The result, of course, is a runaway vehicle. If the driver keeps cool, when the brakes fail, his vehicle can generally be brought back under control.

causes

There are four common causes of brake failure: Loss of air pressure; air blockage; brake fade; mechanical failure.

First and foremost is the loss of air pressure. When the hydraulic system won't build up pressure, the brake pedal will feel spongy and go to the floor.

When air is low, a warning buzzer should sound. You should stop immediately. Further air loss could prevent you from being able to stop.

Your brakes should apply automatically when air loss reaches a critical level leaving you stopped on the road. The brake system is intended to retain, or hold, enough air to stop the vehicle. Be aware that, even when everything works as designed, air loss could be too fast, and air could be exhausted before the vehicle is stopped. Also, the independent trailer brake valve won't activate the trailer brakes, because they depend on the air system also.

On a vehicle equipped with spring loaded parking brakes, the brakes will activate automatically when air pressure fails. This will generally bring the trailer to a stop if it isn't on a steep downgrade.

A blockage can prevent air from reaching the brakes. This is commonly caused by water freezing in the air system. Air lines can come unhooked, causing problems.

On long downgrades, brakes may overheat and fade. The brakes will lose the power it takes to stop wheel rotation. The brakes won't cool quickly.

Brake failure can result from a failure in the mechanical linkage. Rarely are all brakes affected at once. The vehicle can usually be stopped.

procedure

Downshift. Putting the vehicle into a lower gear will help to slow it.

Pump the brakes. Sometimes pumping the brake pedal will generate enough air pressure to stop the vehicle.

Use the parking brake. The parking or emergency brake is separate from the hydraulic brake system. Therefore, it can be used to slow the vehicle. However, be sure to press the release button or pull the release lever at the same time you use the emergency brake so you can adjust the brake pressure and keep the wheels from locking up.

Find an escape route. While slowing the vehicle, look for an escape route — an open field, side street or escape ramp. Turning uphill is a good way to slow and stop the vehicle. Make

experience speaking

"With the new driver, what we're finding is they're braking the vehicle too hard. What we want to do is educate the right foot not only on the throttle, but on the brake pedal as well. Most, if not all, of the students have never driven a vehicle that can bend in the middle, so to speak, such as a tractor-trailer unit. That hard, hard braking is the one thing we want to avoid."

John
Instructor
Wisconsin

experience speaking

"When going downhill and using the brake, a steady brake, an even brake, is most appropriate. Pumping the brake or stabbing the brakes is inappropriate in that it will cause the brakes to overheat and does not really control the speed of the vehicle. It allows the vehicle to change speeds rather rapidly. Whereas, a gentle, even brake application is appropriate. You also want to make sure that if you are in the mountains where you're using your brakes heavily, that you have the brakes adjusted properly, prior to going down the hill. You also want to make sure that you apply enough pressure. You may only have the tractor brakes applied, and the trailer brakes may not be applied as much. That's supposed to be taken care of in the valving, but, in fact, in reality, it does not always happen. So, therefore, you need to apply enough brake so that you have brake applications on all wheels."

Dave
Instructor
Wisconsin

sure the vehicle does not start rolling backward after you stop. Put it in low gear, apply the parking brake, and if necessary roll back into some obstacle that will stop the vehicle.

brake failure on downgrades

Brake failure on downgrades. Going slow enough and braking properly will almost always prevent brake failure on long downgrades. Once the brakes have failed, however, you are going to have to look outside your vehicle for something to stop it.

Your best hope is an escape ramp. If there is one, there'll be signs telling you about it. Use it. Ramps are usually located a few miles from the top of the downgrade. Every year, hundreds of drivers avoid injury to themselves or damage to their vehicles by using escape ramps. Some escape ramps use soft gravel that resists the motion of the vehicle and brings it to a stop. Others turn uphill, using the hill to stop the vehicle and soft gravel to hold it in place.

Any driver who loses brakes going downhill should use an escape ramp if it's available. If you don't use it, your chances of having a serious crash may be much worse.

If no escape ramp is available, take the least hazardous escape route you can, such as an open field, or a side road that flattens out or turns uphill.

Make the move as soon as you know your brakes don't work. The longer you wait, the faster the vehicle will go and the harder it will be to stop.

Blowouts

The term "blowout" refers to the sudden loss of tire inflation. Blowouts can result from worn tires, cracks in the tire casing or from damage from debris, potholes or nails. (Blowouts resulting from wear can be prevented with the thorough pretrip inspection required by law.)

The blowout of a front tire disrupts steering and can cause the vehicle to veer in the direction of the blowout. With a rear tractor tire blowout, you may feel the rear of the tractor slide from side to side, or there may be no symptoms at all. A trailer tire blowout can't be felt, only seen. You might only notice a trailer tire blowout except through your mirrors or the signals of other drivers.

Quickly knowing you have a tire failure will let you have more time to react. Having just a few seconds to remember what it is you're supposed to do can help you. The major signs of tire failure are sound, vibration and feel.

The loud bang of a blowout is an easily recognized sign. Because it can take a few seconds for your vehicle to react, you might think it was some other vehicle, but anytime you hear a tire blow, you'd be safest to assume it was yours.

If the vehicle thumps or vibrates heavily, it may be a sign that one of the tires has gone flat. With a rear tire, that could be the only sign you get.

If the steering feels heavy, it's probably a sign that one of the front tires has failed. Sometimes, failure of a rear tire will cause the vehicle to slide back and forth or fishtail. However, dual rear tires usually prevent this.

procedure

There are four important things that safe drivers do to handle a tire failure safely: be aware that a tire has failed; hold the steering wheel firmly; stay off the brake; after stopping, check all the tires.

If a front tire fails, it can twist the steering wheel out of your hand. The only way to prevent this is to keep a firm grip on the steering wheel with both hands at all times. Hit the throttle to the floor as this will help you to maintain steering control.

It's natural to want to brake in an emergency. However, braking when a tire has failed could cause loss of control. Unless you're about to run into something, stay off the brake until the vehicle has slowed down. Then brake very gently, pull off the road, and stop.

After you've come to a stop, get out and check all the tires. Do this even if the vehicle seems to be handling alright. If one of your dual tires goes, the only way you may know it is by getting out and looking at it.

emergency maneuvers **review**

1. Stopping is not always the safest thing to do in an emergency: True or False?

2. What are some advantages of going right instead of left around an obstacle?

3. If a tire blows out, you should put the brakes on hard and downshift to stop quickly: True or False?

4. What two things are important to prevent rollover?

5. When you turn suddenly while pulling doubles, which trailer is most likely to turn over?

6. Why should you not use the trailer hand brake to straighten out a jackknifing trailer?

7. What is offtracking?

8. What is evasive steering?

9. Describe the best procedure for offroad recovery.

10. What are four common causes of brake failure?

Skid Control & Recovery 6 – 9

causes of skids

tractor/trailer skids

skid recovery

Skid Control & Recovery 6 – 9

A skidding vehicle is a vehicle out of control.

The first and major cause of all skids is excessive speed for conditions. Every driver has the responsibility to maintain a constant awareness of road and weather conditions and to adjust speed accordingly.

Skidding can lead to running off the road, rolling over and collisions. In many emergencies, skidding prevents drivers from taking counter measures.

Friction (traction), wheel load and the forces of motion are the factors that affect vehicle control. Skids happen when there is an imbalance among vehicle control factors.

Skids occur while braking when the forces of motion are greater than the vehicle's traction or braking power. They occur while steering when the forces of motion are greater than the traction of the turning tires. In the latter case, the vehicle continues in the direction it was moving rather than turning. A skid happens whenever the tires lose their grip on the road.

Causes of Skids

Overbraking is braking too hard and locking up the wheels. Skids also can occur when using the speed retarder when the road is slippery.

Oversteering is turning the wheels more sharply than the vehicle can turn.

Overacceleration is supplying too much power to the drive wheels, causing them to spin. Drivers who adjust their driving to conditions don't have to overbrake or oversteer from too much speed.

By far the most common skid is one in which the rear wheels lose traction through excessive braking or acceleration. Skids caused by acceleration usually happen on ice or snow. They can be stopped easily by taking your foot off the accelerator. If it's very slippery, push the clutch in. Otherwise the engine can keep the wheels from rolling freely and regaining traction.

Rear wheel braking skids occur when the rear drive wheels lock. Because locked wheels have less traction than rolling wheels, the rear wheels usually slide sideways in an attempt to catch up with the front wheels. With vehicles towing trailers, a drive-wheel skid can let the trailer push the towing vehicle sideways, causing a sudden jackknife. As always, remember: a locked or spinning wheel wants to lead.

Tractor/Trailer Skids

Tractor-trailer skids are classified based on what happens during the skid. There are four major kinds of skids: trailer jackknife; tractor jackknife; front wheel skid; and all-wheel skid.

trailer jackknife

Overbraking or oversteering can cause a trailer jackknife. Or, both can happen at once, magnifying the potential for a trailer jackknife.

When overbraking, the trailer will continue moving forward at a higher speed than the tractor. Not being able to come forward, it will come around.

When taking a curve too fast for surface conditions, the rear of the trailer may continue in the same direction it was originally heading while the tractor and the front of the trailer turn.

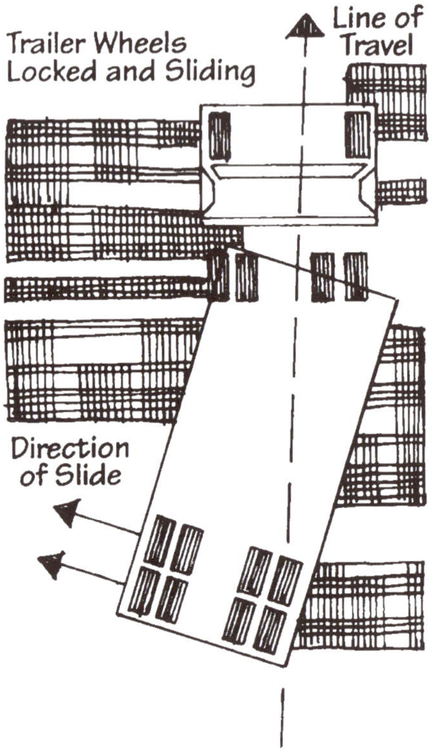

To prevent a trailer jackknife, simply take a few precautions:

Inspect your air system, and check brake adjustment;

Adjust your speed to suit conditions;

Read the road ahead;

Avoid braking in curves;

Avoid hard braking at all times.

tractor jackknife

A tractor jackknife occurs when the drive wheels lose traction. The drive wheels attempt to overtake the **front** wheels. So the rear of the tractor tends to swing out. The tractor follows the path of least resistance while the trailer continues in original direction. Then, as the trailer pivots, it pushes the rear of the tractor outward. The result is a jackknife.

To prevent a tractor jackknife, avoid overbraking, overacceleration and sudden downshifts. Load your cargo properly, and make sure it's secure. Do your required inspection including the brake system and tire tread before you pull out.

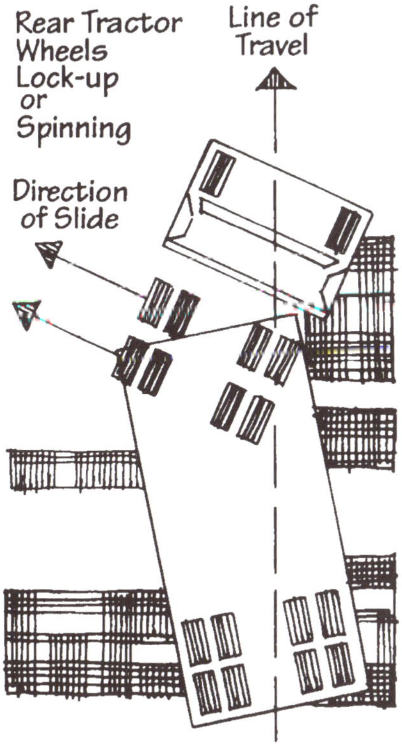

front wheel skids

Front wheel skids are caused by reduced front wheel traction, which can be the result of an excessive load on the fifth wheel, lack of tread on the front tires, hydroplaning, oversteering or brake system malfunction.

The key to preventing front wheel skids begins with a good inspection: Tires, front wheel alignment, suspension system, fifth wheel lubrication.

Reduce your speed on wet or slippery pavement.

all-wheel skids

In an all-wheel skid, all wheels are locked. Friction (traction) changes from rolling to sliding. The wheels can't turn. Usually, during an all-wheel skid, your rig continues in a straight line in spite of any effort to steer. The main cause is overbraking on slippery surfaces.

Avoid excessive braking, especially on slippery surfaces. If it's necessary to brake when a skid is likely or in process, use stab braking to maintain control.

Skid Recovery

Stop braking or accelerating. Push clutch in and steer. This will let the rear wheels roll again, and keep them from sliding any further. If you're on ice, push in the clutch to let the wheels turn freely.

Turn quickly. When a vehicle begins to slide sideways, quickly steer in the direction you want the vehicle to go — down the road. You must turn the wheel quickly.

Countersteer. As your tractor turns back on course, it has a tendency to keep turning. Unless you turn the steering wheel quickly the other way, you may find yourself skidding in the opposite direction.

Learning to stay off the brake, turning the steering wheel

quickly, pushing in the clutch and countersteering in a skid takes a lot of practice. The best place to get this practice is on a driving range or skid pad.

front wheel skids

Most front wheel skids are caused by driving too fast for conditions. Other causes are lack of tread on the front tires and cargo loaded so not enough weight is on the front axle. In a front wheel skid, the front end tends to go in a straight line regardless of how much you turn the steering wheel. On a very slippery surface, you may not be able to steer around a curve or turn.

The interaxle differential on a twin screw tandem can cause the front axle to not want to turn. On slippery surfaces, the interaxle differential should not be locked.

When a front wheel skid occurs, the only way to stop it is to let the vehicle slow down. Stop turning and braking so hard. Slow down as quickly as possible without skidding.

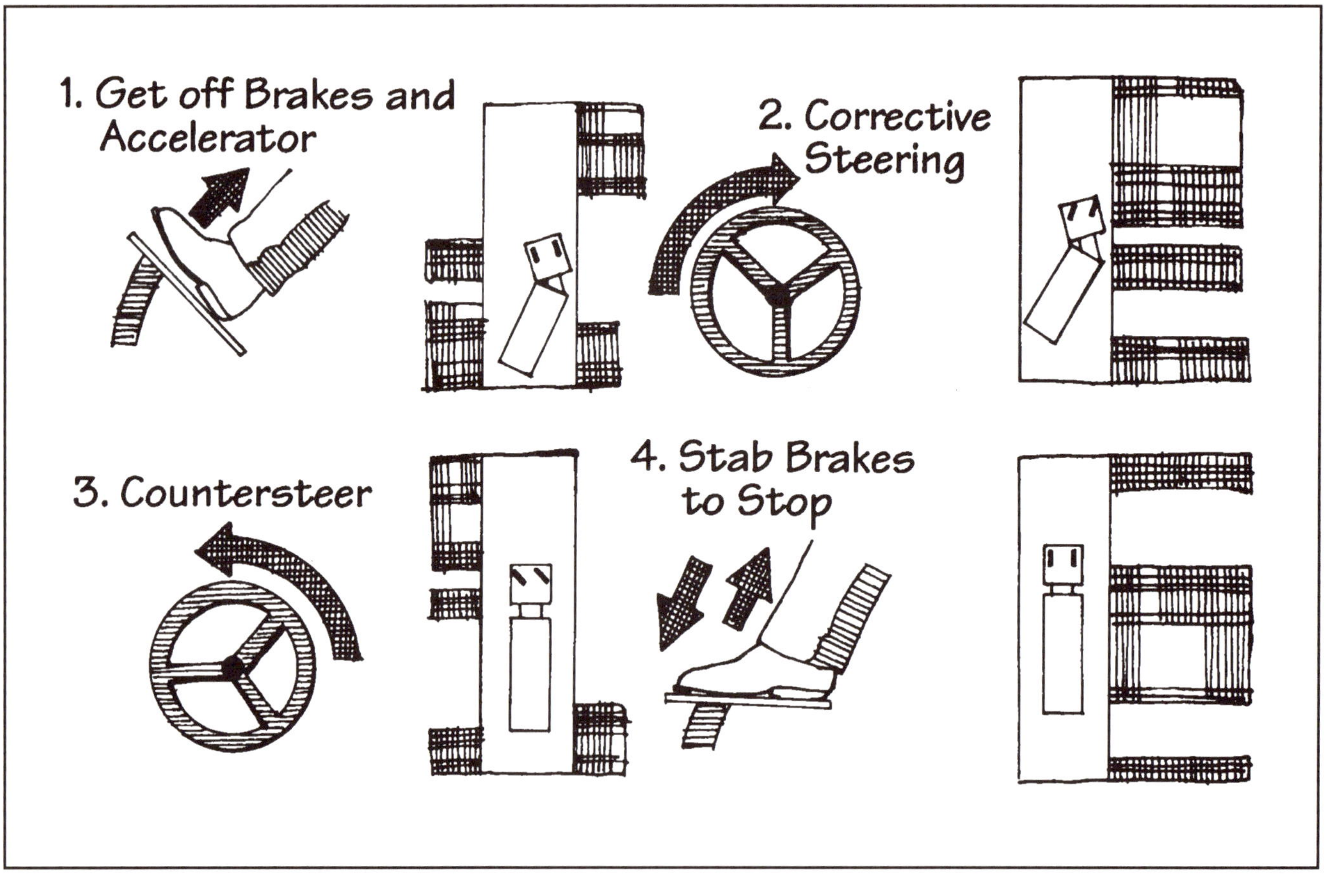

skid control & recovery review

1. A skidding vehicle is a vehicle out of control, if you didn't plan to skid: True or False?

2. What's the leading cause of skids?

3. What's the difference between a tractor jackknife and a trailer jackknife?

4. What's the difference between a front wheel skid and an all-wheel skid?

5. What's the only way to stop a front wheel skid?

6. The most common skid is one in which the rear wheels lose traction through excessive braking or acceleration: True or False?

Transporting Cargo 7

- inspecting cargo
- weight & balance
- securing cargo
- tank vehicles
- other cargo needing special attention
- cargo documentation
- freight pickup
- freight delivery
- special pickup & delivery situations
- cargo handling problems & consequences

Transporting Cargo 7

All drivers must understand some basic cargo safety rules to get a Commercial Driver's License, because the CDL allows the driver to transport cargo in a truck.

If you load cargo wrong, it can be a danger to others and to yourself. Other highway users can hit or be hit by loose cargo. Your vehicle can be damaged by overload. Steering can be affected badly if you don't load cargo right. Loose cargo can hurt you during a quick stop or crash.

Whether or not you load and secure the cargo yourself, you're responsible for inspecting your cargo, recognizing overloads and poorly balanced weight, and knowing your cargo is securely tied down or covered.

If you intend to carry hazardous material that requires placards on your vehicle, you'll also have to have a hazardous materials endorsement to your CDL.

Inspecting Cargo

As part of your pretrip inspection, check for overloads, poorly balanced weight, and cargo that isn't secured right.

Inspect the cargo and its securing devices again within 25 miles after beginning a trip. Make any needed adjustments.

Check the cargo and securing devices as often as necessary during a trip to keep the load secure. Inspect again after you've driven for three hours or 150 miles and after every break you take during driving.

Federal, state and local regulations concerning weight, securement, cover and truck routes vary greatly from place to place. Know the regulations where you'll be driving.

Weight & Balance

definitions

You're responsible for not being overloaded. Following are some definitions of weight measurements you should know.

Gross Vehicle Weight (GVW): The total weight of a single vehicle, plus its load.

Gross Combination Weight (GCW): The total weight of a powered unit, plus trailer(s), plus the cargo.

Gross Vehicle Weight Rating (GVWR): The maximum GVW specified by the manufacturer for a single vehicle, plus its load.

Gross Combination Weight Rating (GCWR): The maximum GCW specified by the manufacturer for a specific combination of vehicles, plus its load.

Axle Weight: The weight transmitted to the ground by one axle or one set of axles.

Tire Load: The maximum safe weight a tire can carry at a specified pressure. This rating is stated on the side of each tire.

Suspension Systems: Suspension systems have a manufacturer's weight capacity rating.

Coupling Device Capacity: Coupling devices are rated for the maximum weight they can pull or carry.

legal weight limits

You must keep weights within legal limits. States have maximums for GVWs, GCWs and axle weights. Often, maximum axle weights are set by a bridge formula. A bridge formula permits less maximum axle weight for axles that are closer together. This is to prevent overloading bridges and roadways.

Overloading can have bad effects on steering, braking and speed control. Overloaded trucks have to go very slow on upgrades. Worse, they may gain too much speed on downgrades. Stopping distance increases. Brakes can fail when forced to work too hard.

During bad weather or in mountains, it may not be safe to operate at legal maximum weights. Take this into account before driving.

top-heavy loads

The height of the vehicle's center of gravity is very important for safe handling. A high center of gravity (cargo piled up high or heavy cargo on top) means you're more likely to tip over. It's most dangerous in curves or if you have to swerve to avoid a hazard. It's very important to distribute the cargo so it's as low as possible. Load with the heaviest cargo on the bottom and lightest on top.

balance the weight

Poor weight distribution can make vehicle handling unsafe. Too much weight on the steering axle can cause hard steering. It can damage the steering axle and tires. Underloaded front axles (caused by shifting weight too far to the rear) can make the steering axle weight too light to steer safely. Too little weight on the driving axles can cause poor traction. The drive wheels may spin easily. During bad weather, the truck may not be able to keep going. Weight that's loaded so there's a high center of gravity causes greater chance of rollover. On flatbed vehicles, there's also a greater chance that the load will shift to the side or fall off.

Securing Cargo

Blocking is used in the front, back or sides of a piece of cargo to keep it from sliding. Blocking is shaped to fit snugly against cargo. It's secured to the cargo deck to prevent cargo movement. Bracing is also used to prevent movement of cargo. Bracing goes from the upper part of the cargo to the floor and/or walls of the cargo compartment.

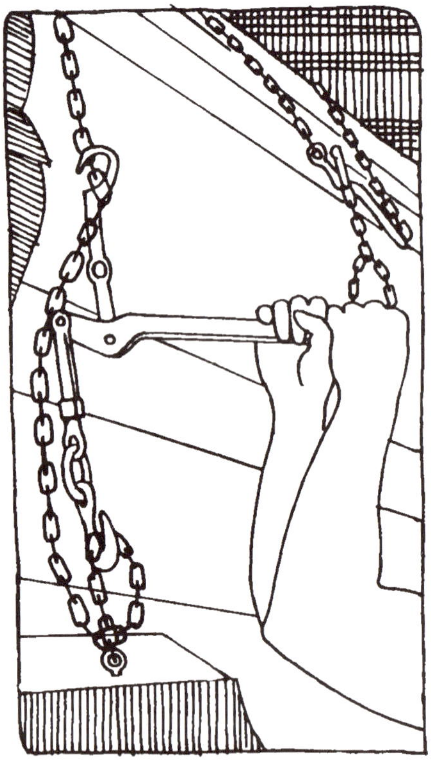

cargo tiedown

On flatbed trailers or trailers without sides, cargo must be secured to keep it from shifting and falling off. In closed vans, tiedowns can also be important to prevent cargo shifting that may affect the handling of the vehicle. Tiedowns must be of the proper type and proper strength. The combined strength of all cargo tiedowns must be strong enough to lift one and one half times the weight of the piece of cargo tied down. Proper tiedown equipment must be used, including ropes, straps, chains and tensioning devices (winches, ratchets, clinching components). Tiedowns must be attached to the vehicle correctly (hook, bolt, rails, rings).

Cargo should have at least one tiedown for each 10 feet of cargo. Make sure you have enough tiedowns to meet this need. No matter how small the cargo is, it should have at least two tiedowns holding it.

There are special requirements for securing various heavy pieces of metal. Find out what they are if you're going to carry such loads.

header boards

Front-end header boards (headache racks) protect you from your cargo in case of a collision. Make sure the front-end structure is in good condition. The front-end structure should block the forward movement of any cargo you carry.

covering cargo

There are two basic reasons for covering cargo: To protect people from spilled cargo and to protect the cargo from weather. Spill protection is a safety requirement in many states. Drivers need to be familiar with the laws in the states they drive in.

You should look at your cargo covers in the mirrors from time to time while driving. A flapping cover can tear loose, uncovering the cargo, and possibly blocking your view or someone else's.

sealed & containerized loads

Containerized loads are generally used when freight is carried part way by rail or ship. Delivery by truck occurs at the beginning or end of the journey.

Some containers have their own tiedown devices or locks that attach directly to a special frame. Others have to be loaded onto flatbed trailers. They're secured with tiedowns just like any other large cargo.

Tank Vehicles

A tank vehicle is a vehicle used to transport any liquid or liquefied gaseous material in a permanently attached tank or a portable tank having a capacity of 1,000 gallons or more. Hauling liquids in tanks requires special skills, because of high center of gravity and liquid movement.

high center of gravity

High center of gravity means that much of the load's weight is carried high up off the road. This makes the vehicle top-heavy and an easy victim of roll over. Liquid tankers are especially easy to roll over. Tests have shown that tankers can turn over at the speed limits posted for curves. Take highway curves or on and off ramp curves well below the posted speeds.

danger of surge

Liquid surge results from movement of the liquid in partially filled tanks. This movement can have bad effects on handling. For example, when coming to a stop, the liquid will surge back and forth. When the wave hits the end of the tank, it tends to push the truck in the direction the wave is moving. If the truck is on a slippery surface, such as ice, the wave can shove a

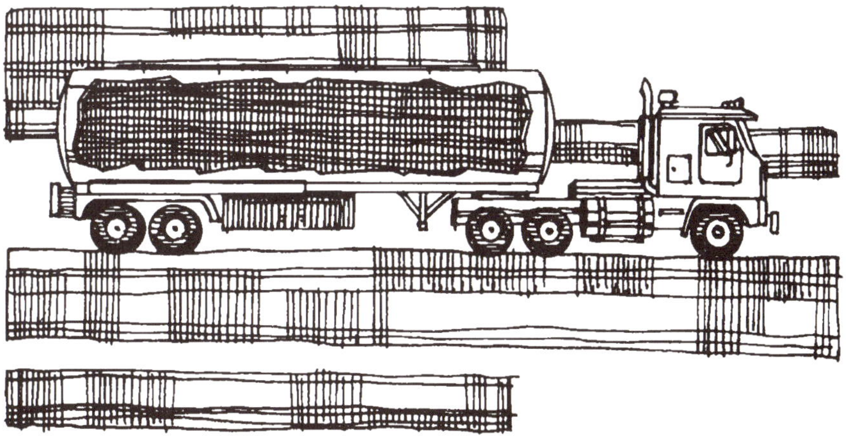

stopped truck out into an intersection. The driver of a liquid tanker must be very familiar with the handling of such vehicles.

bulkheads

Some liquid tanks are divided into several smaller tanks by bulkheads. When loading and unloading the smaller tanks, the driver must pay attention to weight distribution. Don't put too much weight on the front or rear of the vehicle.

baffles

Baffled liquid tanks have bulkheads with holes in them that let liquid flow through. Baffles help to control forward and backward liquid surge. However, side-to-side surge can still occur which can cause a roll over. Be extremely cautious (slow and careful) in taking curves or making short turns with a partially or fully loaded liquid tanker.

smooth bore tanks

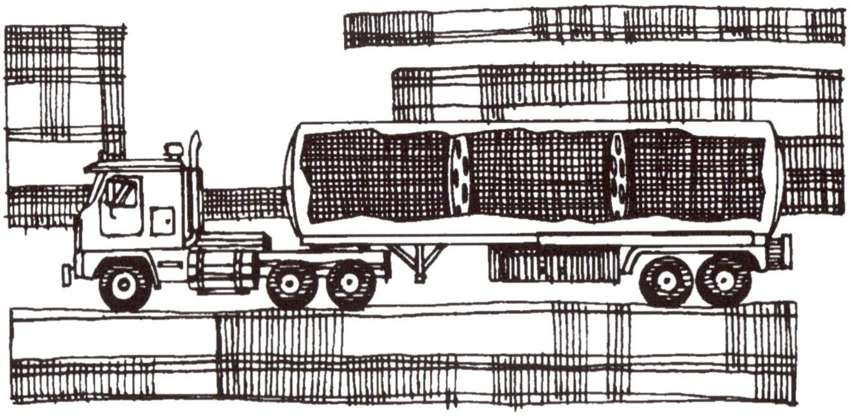

Unbaffled liquid tankers (sometimes called smooth bore tanks) have nothing inside to slow down the flow of liquid. Therefore, forward-and-back surge is very strong. Unbaffled tanks are usually those that transport food products (milk, for example). (Sanitation regulations forbid the use of baffles because of the difficulty in cleaning the inside of the tank.) Be extremely cautious (slow and careful) in driving smooth bore tanks, especially when starting and stopping.

outage

Never load a cargo tank to completely full. Liquids expand as they warm, and you must leave room for the expanding liquid. This is called outage. Since different liquids expand by different amounts, they require different amounts of outage. You must know the outage requirement when hauling liquids in bulk.

dense liquids

A full tank of very dense liquid (such as some acids) may exceed legal weight limits. For that reason you may often only partially fill tanks with heavy liquids. The amount of liquid to load into a tank depends on the amount the liquid will expand in transit, the weight of the liquid and legal weight limits.

Other Cargo Needing Special Attention

dry bulk

Dry bulk tanks require special care because they often have a high center of gravity, and the load can shift. Be extremely cautious going around curves and making sharp turns.

hanging meat

Hanging meat (suspended beef, pork, lamb) in a refrigerated truck can be a very unstable load with a high center of gravity. Particular caution is needed on sharp curves such as on and off ramps. Go slow.

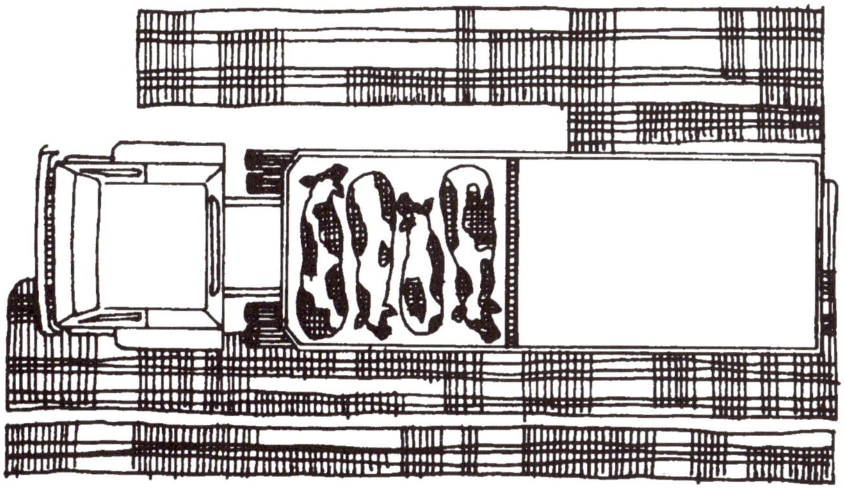

livestock

Livestock can move around in a trailer, causing unsafe handling. With less than a full load, use false bulkheads to keep livestock bunched together. Even when bunched, special care is

experience speaking

"A bill of lading is a contract. That's the contract between this company and the shipper. It has to be completed. Everything on that bill of lading has to be filled out. Where we have agents that can do it for the driver, that's fine. But our drivers often act as our agent. They are the last person to deal with that shipper when they deliver the goods. The shipper has to receive his copy at origin, and he has to get a copy of it at destination. Everything on that contract has to be filled out, because it is a legal document. So, therefore it has to be signed by the driver. It has to be signed by the shipper. All the areas have to be filled out, everything has to be filled out."

Roberta
Safety Director
Florida

necessary because livestock can lean on curves. This shifts the center of gravity and makes rollover more likely.

oversized loads

Overlength, overwidth and overweight loads require special transit permits. Driving is usually limited to certain times. Special equipment may be necessary such as wide-load signs, flashing lights and flags. Such loads may require a police escort or pilot vehicles bearing warning signs or flashing lights. These special loads require special driving care. As with other loads covered here, be extremely cautious (slow and careful).

Cargo Documentation

There are essentially two types of motor carriers: private carriers and for-hire carriers. Cargo documentation varies somewhat from one type of carrier to the next.

Private carriers transport their own goods and products. Soft drink bottling companies, bakeries and meat packing plants are just three of many common examples. Since private carriers transport their own products or raw materials, legal documentation is generally not required, but company documentation (such as receipts and scale tickets) usually is.

For-hire carriers transport goods or products owned by others for a fee. There are three types of for-hire carriers: Common carriers; contract carriers; and exempt commodity carriers.

Common carriers are required to be licensed by the Interstate Commerce Commission (ICC) to carry specific commodities between one area and another.

Contract carriers also get operating rights from the ICC to haul commodities under contract for a specific shipper. Exempt commodity carriers transport non-regulated, or exempt, commodities (usually agricultural or seafood). Since for-hire carriers haul goods for others, legal documentation is required.

If you are employed in a business which requires freight documentation (the legal papers), you must learn what documents are required and what your responsibility is in the handling of them. Every detail can't be covered here: It's your responsibility to learn specifically what's required when you are employed by a motor carrier.

definitions

You're responsible for certain documentation depending on the type of carrier and the cargo being transported. Following are some definitions you should know.

Carrier: An individual or company in the business of transporting goods.

Shipper (Consignor): Individual or company originating the order for transport of goods.

Receiver (Consignee): The person or company to whom the goods are shipped, or consigned to.

Freight Broker: A person or company who arranges transportation of goods.

Freight Forwarder: Company that puts together small shipments from various shippers into larger shipments. Ships to break bulk stations where shipment is broken down and sent to proper destinations.

Originating Carrier (Pickup Carrier): The carrier who picks up a shipment from a shipper.

Connecting Carrier: The carrier who delivers shipment to interchange point where goods are transferred to another company to continue shipment.

Terminal Carrier (Agent): The carrier who delivers the shipment to the consignee.

Bill Of Lading: Written transportation contract between shipper and carrier. Identifies freight, who it's consigned to, place of delivery and terms of agreement.

Straight Bill Of Lading: Provides that goods on bill be delivered to consignee indicated. Consignee doesn't have to surrender his copy to get goods.

Order Notify Bill Of Lading: Negotiable document which enables shipper to collect before shipment reaches destination. Indicated consignee must surrender his copy of bill of lading to get shipment.

Through Bill Of Lading: Covers shipment by more than one carrier at a fixed rate for entire service.

Manifest: Document describing contents of an entire shipment on a vehicle.

Packing Slip: Detailed list of packed goods prepared by the shipper.

experience speaking

"Well, the bill of lading is a legal contract. When the driver signs that bill of lading, he signs for the company that everything that's represented on the bill is there. If he doesn't count his freight, or if he doesn't inspect his freight, he may end up losing himself and the company a lot of money."

Tom
Safety Director
Florida

experience speaking

"When it comes down to court proceedings where we're trying to get money that is due us, or if the shipper wants a refund back for something he really shouldn't get a refund for, if he hasn't signed that contract, we have nothing to stand on. We have no legal document."

Roberta
Safety Director
Florida

Delivery Receipt: Receipt signed by customer receiving goods to indicate acceptance of goods from the driver.

Warehousing Receipt: Receipt for goods placed in a warehouse.

Charges: Both goods and transportation services have to be paid for. Payment for goods is usually arranged between shipper and consignee; sometimes the carrier has to collect. Carrier arranges for payment of transportation services. The driver can't deliver or relinquish possession of freight unless charges are paid or arrangements made for payment, including any payments carrier collects for the shipper.

Prepaid: Transportation charges paid or will be paid at shipping point.

COD: Payment for goods is made at delivery point. Driver must know company policy and procedures for COD.

Order Notify Shipment: Payment for shipment of goods on Order Notify Bill of Lading. Driver must collect copy of bill of lading from person receiving goods as payment for goods.

Basis Of Rates: Rates are charges

Per unit of weight (pound, for example)
Per unit of distance (mile, for example)
Rate x weight and distance = charges

Rates are based on value of cargo and services performed by carrier.

Value Of Cargo: Actual valuation -- actual value of goods shown on bill of lading when rate applied is dependent on that fact. Released valuation – value of goods set by shipper as limits of carrier liability and as basis of rates charged.

Weight And Distance: The primary determiner of shipping charges. Rate expressed in terms of per pound of cargo and per mile of travel.

Services Performed By Carrier: List of services that a common carrier performs for the public and the rates for those services.

Tariffs: Rates on transportation charges based on type of service.

Minimum Rate: Lowest lawful rate that may be charged. Each truck has minimum truckload weight which is quantity of freight necessary to qualify for shipment.

Maximum Rate: Highest rate that can be charged for a specific type of shipment. Examples of when maximum rate might be used are less than truckload (LTL) shipments with exclusive use of truck, mixed truckload in single shipment by one shipper, expedited services or redelivery.

Duties: Government tax on imports and exports which must be collected will be indicated on bill of lading. Often these goods are in-bond: Goods on which duty or tax is due; carried by bonded carrier which guarantees duty will be paid. Driver must know procedures for collecting duties for in-bond cargo.

Services And Surcharges: Basic rates, services and duties are determined in advance by rules and agreements between shipper and carrier. Based on such things as type of cargo (valuation), weight, destination, type of service, (LTL, exclusive use of truck, for example). Additional services and rates may be specified on bill of lading. Driver must know what these mean and must perform service as required.

Inside Delivery And Tailgate Delivery: Bill of lading indicates whether freight is to be delivered inside or picked up at tailgate. Charges differ, driver must perform as specified.

Helper Service: Bill of lading will specify who will provide and pay for helpers, if any. Also called "lumpers."

Residential Delivery: Bill of lading will specify if delivery is to be made at a residential address, conditions of delivery and method of collecting payment.

Dunnage And Return Dunnage: Dunnage is cardboard, lumber or other material used to stabilize a shipment. It is not packaging. Weight of dunnage usually specified separately on bill of lading. Usually no charge for dunnage weight. If return dunnage specified, driver must retain dunnage and return to carrier or shipper as indicated.

Storage And Delay Charges: Additional charges are made for storage or if carrier is delayed from making delivery.

Demurrage: Means detention of vehicle containing freight beyond a stipulated time. Payment made to carrier for such a delay. (Also called detention time.)

Freight Pickup

When you sign a bill of lading, it means you acknowledge that your company accepts the contract terms. Since you are acting as your company's agent, your signature legally binds the company. Your signature on the bill of lading means you accept responsibility for the freight; that the actual freight matches that on the bill of lading in kind and amount.

You must count, inspect and generally determine that the freight is properly described before signing the bill of lading.

Freight must be accurately and properly marked. You should refuse or note in writing if freight is short, damaged, poorly packed or poorly loaded. You can do this on the bill of lading.

Freight Delivery

There are several key points to keep in mind when delivering freight. Make sure your delivery is made to the right person. Handle payment for merchandise and freight charges correctly. Get the proper signatures on the freight bill and bill of lading. Follow to the letter your company's policy concerning short shipments, damages or any other problems. Remember, the freight is yours until the consignee signs for it.

Upon delivery of freight you may have to do the unloading, or others may be assigned to unload. Either way, you, as the driver, are responsible for the freight until it's signed for. If you don't unload yourself, you must still observe and supervise.

Special Pickup & Delivery Situations

split shipments

A split shipment is one in which your cargo is picked up at more than one address. A receipted bill of lading must be furnished at each address. The bill of lading must also contain the address for each delivery if the shipment is going to several locations.

Be especially careful with split shipments. It's easy to get mixed up. Be sure to check each piece against the freight bill very carefully. The wrong carton could easily end up in the wrong city if the driver isn't careful.

interline freight

Interline freight is freight picked up from or delivered to another carrier. With interline freight, you must use the same care (perhaps even more) as if you're picking up from a shipper.

Check carefully for any cargo shortage or damage before signing the connecting carrier's freight bill. Don't sign the freight bill if the connecting carrier refuses to list or acknowledge proper exceptions such as shortages or damages.

Sometimes equipment interchange is involved. This entails you leaving your company's trailer – loaded with freight – with the connecting carrier to be hauled to its ultimate destination. Or it involves you going to another carrier's terminal to hook up to their trailer for delivery at the destination.

Special rules apply with interline freight. Among them is a trailer inspection (including contents) before possession is signed for. Always check with your supervisor for instructions before becoming involved with an interline freight movement.

Cargo Handling Problems & Consequences

Remember, by signing a bill of lading, you become responsible in all ways for your cargo. It is your job to deliver the goods you picked up in the same amount and in the same good shape as you received it, to its intended destination. Each shipment needs to be carefully checked and inspected during all pickup

experience speaking

"It was Thanksgiving day when we had the driver coming through the location to spend the holiday with his family. We made the mistake of authorizing parking the vehicle away from our location. At that point the vehicle was stolen over the holidays. It was broken into. The thieves obviously knew what they were doing and stole the whole load. We were fortunate to recover the equipment and the load intact, with minor damage to the console, where they had started the truck and taken off. We've since changed our procedures on allowing equipment to be parked away from our plant location."

Alan
Safety Director
Kentucky

and delivery situations unless it is a sealed load.

Millions of dollars a year are lost when freight is damaged, lost or stolen. There is an inconvenience, as well as potential loss, to shippers and their customers. There are insurance claims and often devastating lawsuits to contend with. This can all lead to loss of business to carriers and, most directly, job loss for drivers.

Theft, ranging from a stolen carton at the loading dock to the hijacking of an entire tractor-trailer unit at gunpoint, is a leading cause of cargo loss. Whether cargo is lost due to theft, carelessness or even misrouting, the responsibility, including the blame, will often fall on the driver. In some cases, the driver will be lucky if the only consequence of loss or damage is that pay is docked.

You must know and comply with your company's security requirements. There are also general measures that can almost always be taken.

Trailers should be parked back to back when sitting loaded. This makes it difficult for thieves to open doors. Vehicles should be secured. Trailers should never be left unsecured when they're unattended. They should be locked and sealed before you walk away from them. If kingpin locks are provided, use them. They prevent engagement of the fifth wheel.

Clearly, there is much more to being a successful and professional tractor-trailer driver than being able to operate the vehicle.

transporting cargo review

1. For what three things related to cargo are drivers responsible?

2. How often must you stop while on the road to check your cargo?

3. How is Gross Combination Weight Rating different from Gross Combination Weight?

4. Name two situations where legal maximum weights may not be safe.

5. What can happen if you don't have enough weight on the front axle?

6. What is the minimum number of tiedowns for any flat bed load?

7. What is the minimum number of tiedowns for a 20 ft. load?

8. What must you check before transporting a sealed load?

9. How is a bulkhead different from a baffle?

10. Should a tanker with liquid take freeway onramps and offramps at the posted speed limit?

11. Are smooth bore tankers different to drive than those with baffles?

12. What three things determine how much liquid you can load?

Transporting Hazardous Materials 8

what are hazardous materials?

why are there rules?

Transporting Hazardous Materials 8

All drivers should know something about hazardous materials.

You must be able to recognize hazardous cargo, and you must know whether or not you can haul it without having a hazardous materials endorsement to your Commercial Driver's License (CDL).

The material provided here is intended only to provide an overview. If you'll actually be hauling hazardous materials, you must have a hazardous materials endorsement to your CDL. Getting the endorsement requires additional and very specific information and training. This is something you may want to discuss further with your instructor.

What Are Hazardous Materials?

The Federal Hazardous Materials Table names materials that pose a risk to health, safety and property during transportation. You must follow the many rules about transporting them. The intent of the rules is to contain the product, communicate the risk, and ensure safe drivers and equipment.

Why Are There Rules?

Many hazardous products can injure or kill on contact. To protect drivers and others from contact, the rules tell shippers how to package safely. Similar rules tell drivers how to load, transport, and unload bulk tanks. These are containment rules.

The shipper uses a shipping paper and package labels to warn dockworkers and drivers of the risk. Shipping orders, bills of lading and manifests are all examples of shipping papers.

There are many different hazard classes. A material's hazard class reflects the risks associated with it. Shippers indicate the hazard class of hazardous products in the item description on the shipping paper. Similar words show on four-inch diamond (square on point) shaped labels on the containers of hazardous materials. If the label won't fit on the container, shippers put the label on a tag. For example, compressed gas cylinders that will not hold a label will have tags or decals.

After an accident or hazardous material leak, the driver may be unable to speak when help arrives. Fire fighters and police must know the hazards involved in order to prevent more damage or injury. The driver's life, and the lives of others, may depend on quickly finding the shipping papers for hazardous cargo. For that reason, you must tab shipping papers related to hazardous materials, or keep them on top of other shipping papers. You must also keep shipping papers in a pouch on the driver's door, in clear view within reach, or on the driver's seat.

Drivers must use placards to warn others of their hazardous cargo. Placards are signs placed on the outside of a vehicle to show the hazard class(es) of products on board. There are a number of different DOT placards. Like labels, each is turned upright on a point. The person who does the loading must place the placards on the front, rear and both sides of the vehicle.

To ensure safe drivers and equipment, the rules require all drivers of placarded vehicles to learn how to safely load and transport hazardous products. They must have a commercial driver's license with the hazardous materials endorsement.

To get the required endorsement you must pass a written test. You will also need a tanker endorsement if you transport hazardous products in a cargo tank on a truck larger than 26,000 pounds, gross vehicle weight rating.

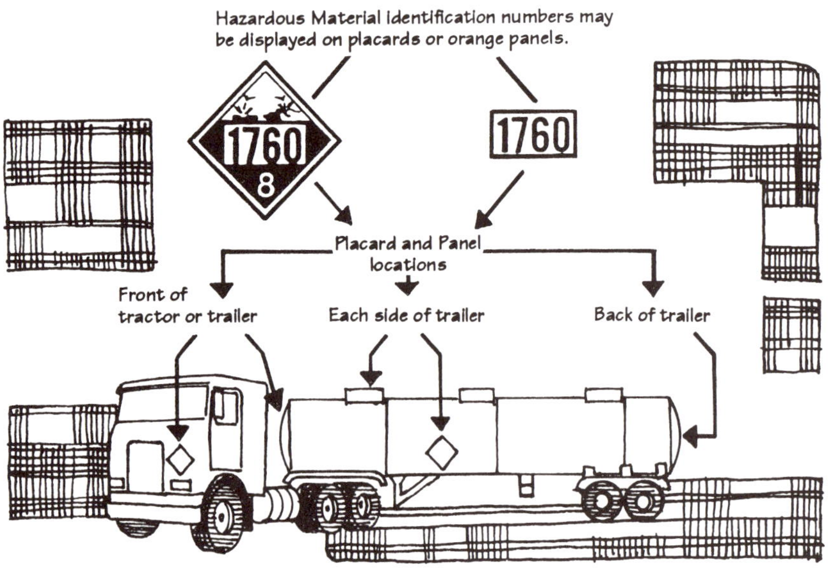

Drivers who need the hazardous materials endorsement must learn the placard rules. If you don't know if your vehicle needs placards, ask your employer. **Never drive a vehicle needing placards unless you have the hazardous materials endorsement. To do so is a crime.** When stopped, you'll be cited and you won't be allowed to drive your truck further. It will cost you time and money. A failure to placard when needed will risk your life and others if you have an accident. Emergency help will not know of your hazardous cargo.

Hazardous materials drivers must also know which products they can load together, and which they can not. Before loading a truck with more than one type of product, you must know if it is safe to load them together. If you don't know, ask your employer.

transporting hazardous materials **review**

1. You must have a hazardous materials endorsement to your CDL to haul hazardous materials. True or False?

2. Where must shipping papers be kept on your vehicle?

3. Give three examples of shipping papers.

4. What is a hazardous materials placard?

5. Why are placards used?

6. What is meant by the term "Hazardous Materials?"

Vehicle Fires 9

causes of fires

how to prevent fires

how to fight fires

choosing a fire extinguisher

Vehicle Fires 9

As many as 500,000 highway vehicle fires occur every year in the United States.

A fire when you're on the road can be a frightening experience. Common sense can do a great deal to prevent many fires altogether and to minimize damage from those that do start.

Causes of Fires

A common cause of fires in vehicles is a short in the electrical system. With more and more electrically-powered equipment on vehicles all the time, these fires have become more of a problem. Electrical fires are often relatively minor in nature, but they can be hard to discover and difficult to fight.

Another common kind of truck fire is tire fires. They're extremely difficult to put out because of the intense heat involved and the likelihood of flare-ups, even long after the fire appears to be out.

More dangerous kinds of vehicle fires can result from accidents such as rollovers and rear-end collisions.

calling for help

If you're involved in an accident that results in a fire, the best thing you can hope for is professional help and fast. If you're not injured and can do so, try your best to reduce the fire danger, but the most important thing to do is get fire fighters to the scene as quickly as possible.

Fire departments need good information to be able to do their job effectively. The more information you can provide, the more specific you can be, the better the response is likely to be.

Describe the location of the fire as completely and exactly as possible in terms of major highway intersections, miles from a specific point, etcetera. Describe the nature of the fire as completely as possible, including a description of the substances involved and the extent of the fire. If fire fighters know what sort of fire they'll be dealing with, they can come prepared with the right supplies and equipment.

experience speaking

"My only experience, personally, with a fire was on a long downgrade in Arizona. A back trailer on a set of doubles had apparently been serviced and the brakes on one side of the trailer were set too tight. The tires didn't catch on fire, but there was enough heat that the grease that was on the wheels from servicing it and some oil did catch fire. The fire extinguishers that are usually in trucks are woefully inadequate in trying to put out a fire of any substance. It took three extinguishers from three different trucks to put that fire out."

Denise
Driver
Nebraska

How to Prevent Fires

Fire needs three elements:

Fuel (a substance that will burn)

Air (oxygen is necessary for combustion to take place)

Heat (a certain temperature must be reached before ignition occurs)

The best way to fight a fire is to prevent its happening at all. Prevention methods all revolve around keeping the three elements necessary for a fire from coming together in a certain spot at a certain time. Common sense, good driving habits and careful housekeeping are the most basic elements of fire prevention.

Vehicle fires occur primarily in three areas of your rig: The cab/engine; tire/brakes; and cargo.

cab/engine fires

Housekeeping is important. Keep the cab clean from paper, oily rags, matchbooks, other flammable refuse. Be sure the maintenance shop keeps the engine clean as excess oil and dirt are invitations to trouble.

Never use gas as a cleaning fluid for anything on your rig or on yourself.

Be sure your maintenance shop makes regular checks of the following items:

Wiring and electrical apparatus (the electrical system is a prime area for fire to start);

Size of fuses used (oversize fuses will allow system overload that may cause a fire);

Battery (should be clean, covered, and all terminal connections should be tight);

Exhaust system (no part of the system should be close to any body parts which are potentially ignitable).

Turn your engine off when you stop to refuel.

Use good sense about smoking. Don't smoke within 50 feet of your vehicle; if you smoke at all, use extreme caution around the equipment and when you're disposing of cigarettes, cigars or matches.

tire and brake fires

Be sure the shop checks brake drums often. Good drums guard against overheating.

Be sure brakes are fully released before you move your rig.

Don't ride your brakes. This habit can cause a fire in the linings that may spread to the tires.

Be sure duals are properly mated and have adequate spacing between pairs.

Check all the tires on your rig at all safety stops. Keep an extra close eye on the inflation of your rig's tires as a simple way to avoid problems with the heat generated by a soft tire. Change soft or flat tires immediately when they're discovered. Thumping a tire won't tell you it's soft, only that it's flat.

Never leave a vehicle with a hot tire until the tire has either been removed or has cooled enough to touch. If there is visible smoldering of the tire, or if you can smell burnt rubber, or if you can't hold the back of your hand against the tire, it's too hot.)

Never place a hot tire in the spare tire rack.

cargo fires

Use your mirrors to keep a lookout for smoke that might be coming from your trailer while you're on the road.

Don't smoke while moving freight in or out of the cargo area or any time you're hauling any kind of hazardous materials.

Know the specific commodities you're hauling so you'll be able to respond correctly and immediately in case of a fire in your load.

Be sure tank grounding lines are attached when loading or unloading tankers to avoid problems with sparking.

How to Fight Fires

No matter how careful you are about keeping your rig clean and well-maintained, you may still find yourself faced with a fire emergency on the road. Don't abandon common sense when you try to fight a fire. Do your best to get the fire under control or put it out entirely, but don't take unnecessary risks that might endanger you or other people who might be nearby. And always report a fire (no matter how small) to your supervisor or terminal manager.

cab/engine

Don't panic! Take rapid stock of the situation you're facing, and begin appropriate action immediately.

The first decision to make is to decide if the fire is one you can handle yourself, or will you need professional assistance? If you need help, use your CB, or have a responsible party call the fire fighters for you.

Check the area quickly for a water supply. If there's usable water nearby and you can move the vehicle nearer to it safely, do so. The capacity of all fire extinguishers is very limited.

Turn the ignition off in a disabled vehicle, and check for any leaking fuel. If you suspect an electrical fire, disconnect the battery cables from the terminals, if you can do so safely.

Don't use water on a petroleum substance fire. It will only make matters worse.

Don't open the hood any farther than necessary to fight an engine fire. The air will cause the fire to grow or spread.

Make the most of your fire extinguisher's capacity. Know the proper way to use it, and then use it wisely.

Keep the wind behind you when you're using an extinguisher to avoid inhaling the fumes and/or smoke from the fire.

Try using sand, dirt or snow to smother the flames if your extinguisher is exhausted and the fire is still not under control, and if there is no water source available.

If a fire appears to be out, don't assume success too quickly. Fires are likely to flare up again. Keep checking the problem area, and never leave the vehicle unattended.

tires and brakes

First make the decision whether you can handle the fire alone or if you'll need help. If you can get a hot tire off safely, do so.

Fire extinguishers that are normally carried on a truck will often not be able to put out a tire fire. They will, however, help to smother surface flames. The best material for fighting tire fires is water. It cools the tire enough to prevent flare-ups. If your extinguisher is empty and no water is available, try throwing dirt or snow on the tire.

cargo

Cargo fires are usually found when you smell smoke or see it escaping around cargo doors. Don't open the doors until the vehicle is in a safe place and fire fighting help is at hand. Even then, extreme caution must be used in opening doors. The powerful effects of a backdraft should be anticipated.

Disconnect the power unit from the trailer, and move it away from the fire area if at all possible. Be sure fire fighters know in advance what your cargo consists of, so they will know what substances they're dealing with.

Prevent excess damage to your cargo by opening doors cautiously and removing freight before the source of the fire is determined, then put out the fire. Avoid indiscriminate use of water on cargo before the source of the fire is located.

If you're dealing with a flammable liquid fire, attack small spill fires from the edges, and try to push them back toward the leaking point. Close covers to extinguish fires in a cargo tank: The air supply will be cut off, and the fire should go out. If you can't get a flammable liquid fire under control, try damming the burning liquid with a dirt dike in order to confine the fire to a smaller area.

If your trailer is refrigerated, you should be aware of the type of refrigerant used. Leakage of some types of agents can cause fire or poison hazard.

Keep other motorists and spectators away from the fire scene. This is especially important in the case of tanker fires. The whole trailer can become a projectile under some conditions.

being a good samaritan

If you stop to assist another driver with a fire, make sure you don't become part of the problem. Park your vehicle far enough away from the fire so it won't be in danger. Try to park upgrade so any leaking fluids won't be able to drain down to your rig and cause a fire hazard.

Be sure you pull off the road completely so you don't create an additional traffic hazard. Setting out emergency warning devices (if they aren't already in place) is a good precautionary measure.

Choosing a Fire Extinguisher

Trucks are susceptible to small fires – carburetor fires, electrical fires, a cigarette smoldering in the upholstery. A little fire can quickly turn into a big one, however, if you don't have the proper fire fighting equipment available.

classes of fires

Four general classifications of fires have been identified by the National Fire Protection Association (NFPA). The fires are grouped according to the type of combustible involved and the type of extinguishing materials needed to fight them.

CLASS A fires occur in ordinary combustible materials like wood, cloth and paper. The most commonly used extinguishing agent is water, but dry chemicals also provide a rapid knockdown of flame and leave a fire retardant film which prevents reflash.

CLASS B fires occur in the vapor-air mixture immediately above the surface of flammable liquids like gasoline, grease and lubricating oils. Dry chemical, carbon dioxide, water, foam and vaporizing liquids can all be used as extinguishing agents.

CLASS C fires occur in electrical equipment where a non-conducting extinguishing agent must be used. Dry chemical, carbon dioxide and vaporizing liquids are suitable. Foam, water and water-type agents conduct electricity, so they should not be used under any circumstances.

CLASS D fires occur in combustible metals such as magnesium, titanium, zirconium and sodium. Normal extinguishing agents generally should not be used on metal fires, because they can increase the intensity of the fire by causing an additional chemical reaction. Specialized techniques and materials are required to extinguish such fires.

types of extinguishers

Standard dry chemical extinguishers (for Class B and C fires only) use a chemical powder to smother the flames. Once the fire is out, the chemical agent can be brushed away easily. Multipurpose extinguishers (Classes A, B, C fires) use a powder that turns to a sticky film on contact with a burning material. This type of extinguisher makes after-fire cleanup much more difficult, but is effective with most kinds of fires.

Gas type extinguishers are either carbon dioxide (CO_2) or halogenated hydrocarbon (Halon). Both types are effective on B and C fires and both are particularly good on electrical fires. Carbon dioxide extinguishers are quite bulky and demand that you get fairly close to the fire. Halon works by suppressing the flames and lowering the temperature of a fire. These extinguishers are less bulky and have a greater range, but are more expensive than carbon dioxide units. Also, Halon is no longer seen as the ultimate weapon in fighting fires that it once was.

A fire extinguisher that's UL (Underwriters' Laboratories) approved, safe to use and effective on at least Class B and C fires, should be a part of your vehicle's basic equipment.

rating fire extinguishers

Fire extinguishers are rated according to the size and type of fires they can put out. A combination of numbers and letters appears on each extinguisher to provide this information. For example: A 5 B:C extinguisher will put out five square feet of a B fire and is effective on electrical fires. A 10 A:40 BC extinguisher will put out ten square feet of a Class A fire or 40 square feet of a Class B fire and is effective on electrical fires.

requirements

Even though common sense should tell you that a fire extinguisher is a necessary item on your vehicle, there's also a requirement in the Federal Motor Carrier Safety Regulations that you carry one 5 B:C (or two 4 B:C extinguishers) on your power unit if you're hauling regular freight, or one 10 B:C extinguisher if you're hauling hazardous materials. The regulations also specify that the extinguisher carried be filled and ready for use, and that it be positioned in an easily accessible location.

experience speaking

"Well, the first thing that they should do is aim the agent at the base of the flame. And from that point work their way to the back edge of the fire of until the flame is out. Once the fire is out, or once they can no longer wee flame, they should shut the extinguisher down. Look again to see if they've missed anything, and attack it from that point, more or less conserving some of the agent. Don't totally dump the unit and then find out something was missed."

James
Instructor
Michigan

fire extinguisher use

You should make yourself familiar with the fire extinguisher(s) carried on the vehicle you drive. Know the type and proper operation. Operation is usually simple and described on the canister.

That seems fairly simple and straightforward. But if you have the opportunity to take some training in fighting fires, you should. It'll give you a chance to practice when there's no real emergency. When you're faced with an actual fire, you don't have time to fumble around with unfamiliar equipment or procedures.

A few basic tips will help you get the most from your fire extinguisher in an emergency. The most important thing to remember is that your extinguisher has a very limited operational time: Many are empty in 8-10 seconds. All the more reason to know what you're doing.

Keep the extinguisher upright. This is a change from old-time extinguishers which needed to be inverted to work properly. Today's pressurized units should not be turned upside down or all the propellant will escape. You must shake dry chemical extinguishers before use.

Attack the fire from the bottom and sweep the nozzle of the extinguisher back and forth in a regular pattern extending at least six inches beyond the flames on all sides. Make the most of your extinguisher's very limited life span.

extinguisher inspection

The fire extinguisher should be part of the routine inspection of equipment on your vehicle. Basic things to look for include:

Does the extinguisher fit snugly in its mounting bracket?

Is the extinguisher properly sealed?

Does the extinguisher have a nozzle on it?

Does the extinguisher gauge show that it's properly charged and ready to go?

Is the extinguisher clean and rust-free?

Does the extinguisher, at the very least, meet the rating requirements of the safety regulations?

If all these requirements are met, you have a fire extinguisher to depend on when you need it. Checks of your fire extinguisher's status should be made at regular intervals for your own protection, as well as that of your vehicle, cargo and the general public.

vehicle fires review

1. Name two causes of fires.

2. What kinds of fires is a B: C extinguisher not good for?

3. When using your extinguisher, should you get as close as possible to the fire?

4. Name some causes of vehicle fires.

5. What three elements are needed for fire to occur?

6. Why are tire fires particularly dangerous?

Accident Procedures 10

warning devices

Accident Procedures 10

There's a standard routine you should follow if you're involved in, or come upon, an accident while you are on the road.

When an accident occurs, your immediate attention is required. Follow this step-by-step procedure at an accident scene.

Step 1: ***Stop immediately!*** Failure to stop is a criminal offense. You're subject to legal penalties (and probably company discipline) if you don't stop after you're involved in an accident (or when you're the first to arrive on the scene.)

Step 2: Do your best to prevent other accidents:

Turn on your four-way flashers for an initial warning signal;

Set out emergency warning devices as required by the Federal Motor Carrier Safety Regulations (FMCSR).

Step 3: If anyone is hurt, do what you can. That may mean administering first aid or arranging for someone else to take care of it, calling for medical help or authorizing someone else to do it. At a minimum, keep body temperature stable and don't move injured persons unless a life-threatening situation exists.

Step 4: Call the state police on your CB or designate someone else to call. As a last resort, go to call yourself if there's no one else present, and your CB is not working. Generally speaking, you shouldn't leave the scene.

Step 5: Get all the information possible down on paper while the accident situation is immediate. Drawing simple diagrams of the situation is helpful.

Step 6: Report full details on the accident to your company as soon as possible. Follow company rules to the letter, and be sure the person you talk to can take care of all the necessary contacts.

Step 7: Be sure a preliminary accident form is completed at the scene. If a representative of your company arrives, he will complete the form. But failing that, you will have to do it yourself. This form is used by your company in its routine investigation of the accident.

Warning Devices

You are required to set out emergency warning devices within 10 minutes of an accident under FMCSR. That means, if you are driving a rig equipped since January 1, 1974, you must have three bi-directional emergency reflective triangles available.

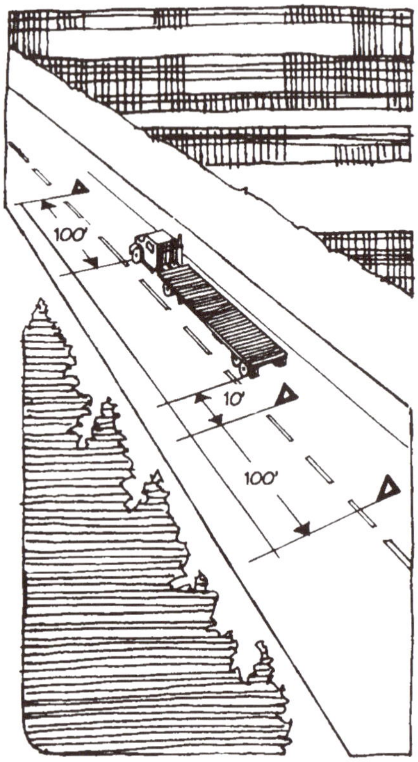

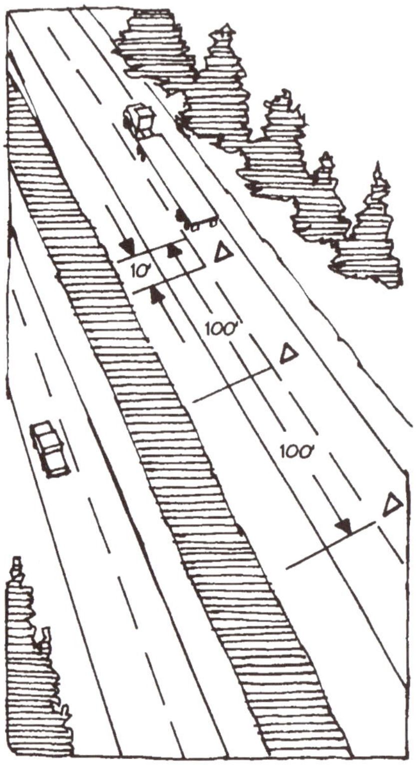

The triangles should be placed in the following locations in the order listed:

At the traffic side of the stopped vehicle, within 10 feet of front/rear depending on traffic direction;

Approximately 100 feet from the stopped vehicle in the center of the lane occupied by that vehicle and toward approaching traffic;

Approximately 100 feet from the stopped vehicle in the opposite direction from the second device.

If your vehicle is older than the 1974 model year, you have several options in emergency warning devices (see Section 393.95 of the FMCSR).

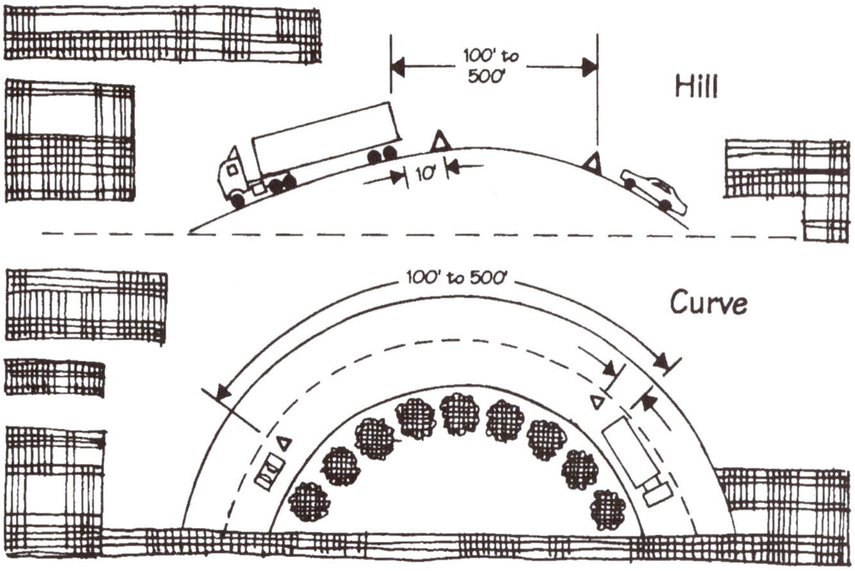

If the accident occurred within 500 feet of a curve, crest of a hill or other obstruction, one warning device should be placed 100 to 500 feet from the stopped vehicle in the direction of the obstruction. This will allow traffic time to react to the hazard. The other two devices should be placed according to rules for two-lane or divided highways, depending on the particular highway or traffic situation.

On divided highways or one-way roads, placement of emergency warning devices is also different. One device should be placed at the traffic side of the vehicle within 10 feet of the rear, another device at a distance of 100 feet and the third device at a distance of 200 feet from vehicle; all in the direction of approaching traffic.

information at the scene

The more detail you can provide in information for your company, the better for insurance or legal purposes later. That means names, license numbers and any other identification information you can obtain on witnesses to the accident.

Be courteous and cooperative in providing information at the scene if requested by a law enforcement officer. But never admit guilt at the scene.

Deciding liability is not up to you!

report promptly to your company

When you report to the company, be specific and include this information:

Exact location of accident;

Precise time of accident;

Reliable estimate of injury/damage;

Location where you can be reached;

Names and addresses of persons involved in the accident;

Names and addresses of insurance companies of all persons involved;

Type, make and model of vehicles involved in accident;

License numbers of all vehicles involved in accident.

Be sure that the person you talk to can take care of contacting the insurance carrier and other necessary parties. If you can't reach your company, contact the nearest office of the insurance carrier yourself, and ask them to contact your company.

experience speaking

"Well I would immediately stop, of course, and see if anyone was hurt. If there was, I would get the proper authorities and medical people out there. Put out my triangles, what have you, and get a law enforcement official there immediately. Fill out the accident form. Get all the witnesses and everything down that I could get. You're going to need that if it gets to court."

Talmage
Driver
Tennessee

a reportable accident

While the FMCSR no longer include requirements for reporting certain accidents, many states and companies do have accident reporting requirements. The actual reporting may be the responsibility of the motor carrier or the driver. But, either way, the information you provide will always have an impact. And there are some unique reporting requirements for hazardous materials accidents.

Obviously an accident should be reported if it involves the death of a human being or bodily injury to a person who requires and receives medical treatment away from the scene. Also, be sure you're aware of the level of property damage that must be present to make an accident reportable in the states you travel through.

unoccupied vehicle

If a vehicle involved in an accident is unoccupied (for example, a parked car), special procedures apply. First, try to find the owner of the vehicle.

If you're unsuccessful in finding the owner of the damaged vehicle, leave your name and address and a phone number where you can be reached. Also leave the name, address and phone number of your company. Be sure it's in a readily seen location; on a sturdy card under the windshield wiper is a good spot.

stay calm

No matter what kind of accident you're involved in, you should remain calm and use your good sense and judgment. If you fall apart in an emergency, you won't be able to help anyone. Take a look at the overall situation, then get started handling the most pressing injuries or problems in the order of priority you mentally establish for handling them.

If a person has an injury that doesn't appear to be life-threatening and you aren't trained in first aid, make the person as comfortable as possible and wait for trained medical personnel.

If the accident victim is in danger of death and you are the only one there, use common sense and do your best to help. If you save a life, you are less likely to be questioned on the first aid technique you used in doing the job.

first aid procedures

You never know when you'll find yourself in a situation where someone needs first aid. It's a good idea to get first aid training through one of the certified courses offered (American Red Cross, for example). If you're faced with injured people and you don't have formal training, you must decide whether to try to help or step back.

Many states have laws which require that the first person to come upon an accident stop and render assistance. If you don't stop, you could be subject to fines or other penalties. This is to say nothing of moral obligations.

You shouldn't be reluctant to offer assistance at an accident. States with "good samaritan" laws offer protection to untrained persons who offer help in good faith in an emergency situation. That means you can't be held civilly liable in those states for the consequences of actions or omissions in rendering emergency care at an accident scene.

Besides your responsibilities involving emergency warning devices, summoning law enforcement officers and emergency medical personnel, you need to assess the extent of injuries to anyone involved in the accident. The following priorities apply in dealing with injured persons:

Is the person breathing? Attend first to anyone who isn't breathing. Be sure the airway is open.

Is the person bleeding? The next thing you should try to deal with is any bleeding, especially if it's severe. Applying direct pressure on the wound is the first step to try.

Is the person in shock? Even if shock symptoms are not yet present in any injured person, it's important to treat for shock in all cases.

Does the person have a fracture? Any fracture will require medical attention as soon as possible.

If you are injured, your responsibilities will obviously be diminished according to the severity of your injuries. One thing to keep in mind if you are injured, but still conscious, is to do your best to assure that your company is notified of the accident. Authorizing someone else to take care of that job may be the best thing to do.

driver responsibility

When an accident happens, your responsibilities as a driver fall into two general areas: Dealing with specific problems at the accident scene (setting out warning devices, calling for help and giving first aid if required) and gathering and reporting complete information on the accident to your company as quickly as possible. Both elements are important. Both should be handled in as professional a manner as possible.

accident procedures review

1. What are some things you can do at an accident scene to prevent another accident?

2. If you come upon the scene of an accident and don't know standard first aid procedures, it's best to keep going in order to keep the scene from becoming overcrowded: True or False?

3. Describe the proper placement of warning devices.

4. Whose responsibility is it to report an accident?

5. What are your responsibilities at the scene of an accident?

6. What information do you need to report to your company concerning an accident?

7. What should you do if you're in an accident with an unoccupied vehicle?

8. If you feel you were responsible for an accident, you should tell a police officer as soon as one arrives at the scene: True or False?

Preventive Maintenance 11

the driver's role in preventive maintenance

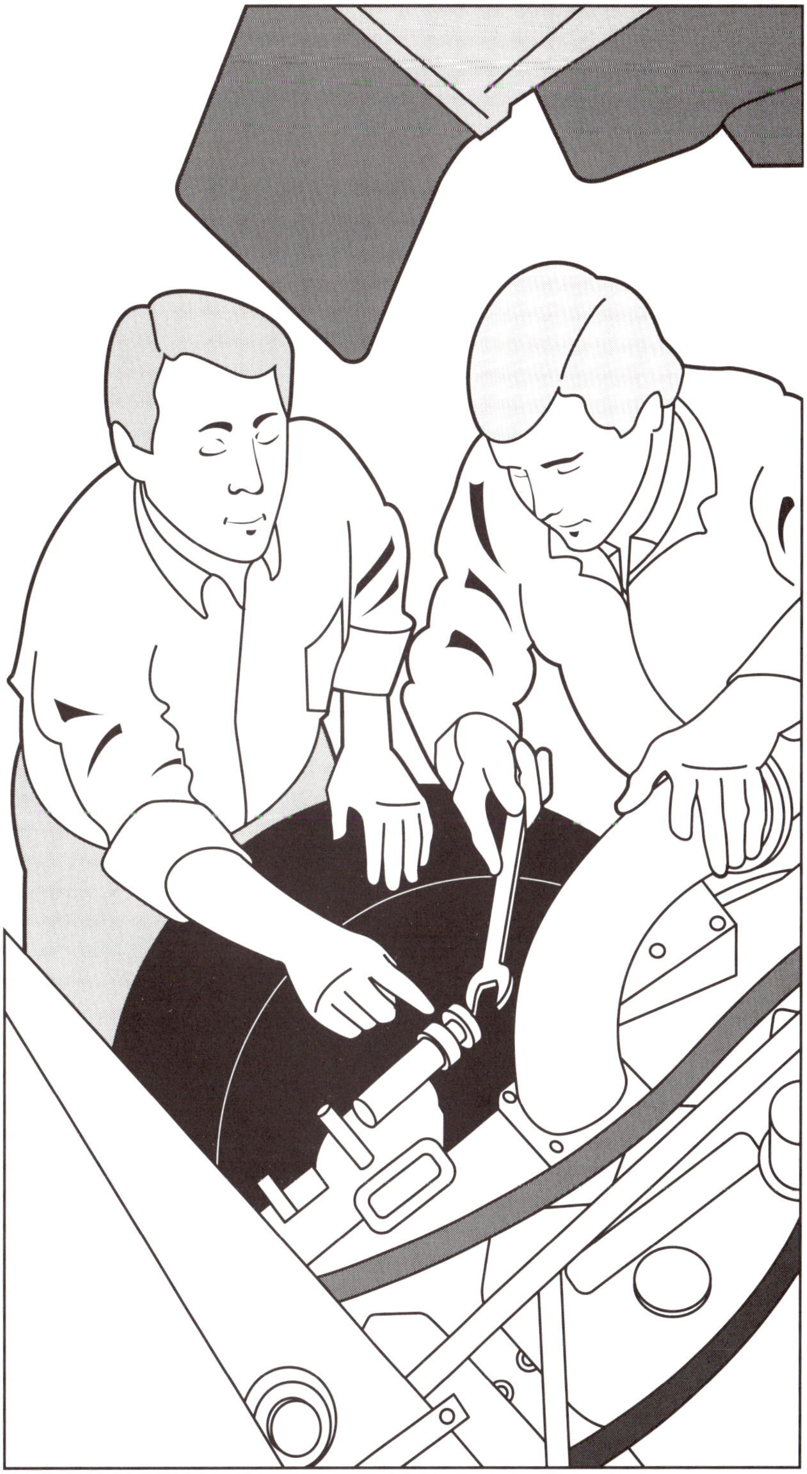

Preventive Maintenance 11

Preventive maintenance is an economical and relatively simple way for a carrier to protect its equipment investment.

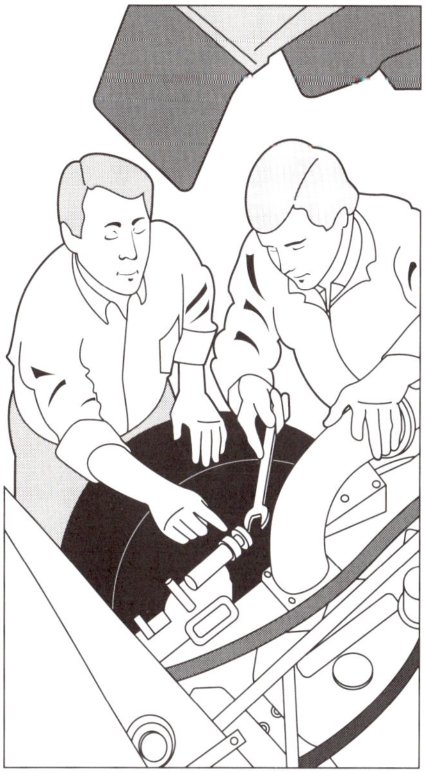

The basic principle of preventive maintenance is to put vehicles on a set maintenance schedule and to routinely replace parts and components on a time or mileage-keyed basis instead of waiting for them to wear out or break down.

Even though preventive maintenance involves a regular outlay of money, it's still cheaper in the long run to do routine replacement and repairs according to a schedule than to wait for breakdowns (which have a habit of occurring at inconvenient times and locations).

The Driver's Role In Preventive Maintenance

More and more fleets are seeing you as a real partner, as the first line of defense, in preventing major maintenance problems. This trend simply recognizes the importance of the role you play in the total operational picture of your vehicle. Your responsibilities can get fairly sophisticated depending on how seriously your carrier takes the whole preventive maintenance approach.

Your powers of observation will be called upon to spot problem situations as early as possible. That on-the-road expertise should work together with the in-shop expertise of the fleet's mechanics.

You're responsible for knowing the mechanical condition of your vehicle and for operating that vehicle correctly and efficiently any time you're behind the wheel. The shop is responsible for providing you with a driveable and safe vehicle.

Do your part to be sure communications are open and friendly between you and the maintenance department. Both sets of personnel are involved in the same effort, keeping your vehicle running in top shape and as efficiently as possible.

experience speaking

"We are adamant about good communication from the driver to our maintenance department. We require a daily inspection. On this inspection sheet, whether it's tractor or trailer, the driver has the opportunity to note the defects, if any, and to further communicate down in the lower area of the report."

John
Safety Manager
Tennessee

Some basic guidelines will help you contribute the most you can to your carrier's preventive maintenance program. Make the complete and careful pretrip inspection of your vehicle that's required before you head out. Treat your rig like it's your own when you're on the road. Don't abuse it in any way by bad driving habits. If you have trouble on the road, follow company procedures for such situations exactly. Know your vehicle well so you can make good judgments about its operation or any defects. Describe any problem you find accurately and in detail, and report it promptly. Aim for good relationships with maintenance personnel. You'll have much better results in getting problems taken care of.

use your senses

Don't forget that you're the only on-the-spot observer of the vehicle you drive and its operation. Your eyes and ears should sense trouble before it gets to be major. Put your senses to work.

Listen for unusual or abnormal equipment sounds. Thumps, squeaks, rattles, bumps, squeals and hisses can announce the beginning of trouble. If things don't sound right, let maintenance know about it.

Smell for unusual odors that signal trouble. Burning rubber, insulation, wood, scorched fabric and hot oil or other fluid can all mean problems. Diagnosis can be made early with the help of a good sense of smell.

Feel changes in the vehicle's response. Steering, braking, shifting, other handling operations all have their unique feel in a particular rig. If the vehicle doesn't behave the way it should, report it promptly. Little problems cost less to fix and cause less downtime.

Observe your equipment carefully in your routine inspections. Defects in lights, wiring, cables, tires, splash guards, locks, air lines, coupling devices, fifth wheels, tarps and fasteners, landing gear, brakes and accessories should all be carefully noted and reported.

driving habits

If you're committed to the idea of preventive maintenance, your commitment will extend to your driving habits. Being a keen observer of your vehicle's behavior isn't the end of your responsibility. Your own driving behavior can get the most out of the equipment for the least cost.

Four major vehicle areas deserve special attention: Tires; brakes; clutch; and engine. In all four, your driving habits can definitely help the preventive maintenance efforts of your fleet.

tires

Heat is the biggest source of tire failure. Heavy loads and high rates of speed compound the problem of heat buildup. How do you minimize the effects of heat on tires? One way is to be sure all tires are inflated to correct levels. Check inflation often with a gauge.

Ideally, tire pressure should be checked when the tire is cold. But if you have to check it when you've been running the vehicle, add about 10-15 psi to the recommended levels for the tires. This pressure increase is normal.

Underinflation is the major problem with tires, but overinflation can negatively affect tire life too. If a tire is more than 10 psi away from the normal pressure for its type, it should be examined for damage, improper size or other possible problems. Repair or adjustment should be made before further use of the tire.

In the case of underinflation, heat is the real culprit. The tire doesn't have even contact with the road surface, and the problem gets worse the faster you go. Besides reducing tread life, excessive heat can cause tread to physically separate from the body and cause breaks in the body cords.

If you run duals with one of the tires underinflated or flat, you have problems too: The good tire winds up carrying an extra load and operating at decreased efficiency. Studies have shown that even a 10 psi difference in pressure between the two tires of a dual assembly can drop performance of the good tire 30-45%.

An overinflated tire doesn't absorb road shocks well. Such a tire is much more susceptible to puncture and other damage from debris on the road and chuckholes. Another problem of overinflation is reduced traction. Tread does not make proper contact with the road surface, and traction is lessened. This situation is especially dangerous during bad weather conditions.

Fast stops and starts can also wear out tires before their time. Hitting debris on the road or chuckholes can damage tires too. Chuckholes can break rims, throw wheels and front ends out of alignment, put tires out of balance, knock off hub caps and sometimes cause complete tire failure. Keep your eyes open and avoid road hazards whenever possible.

Avoid scraping your tires against curbs or other obstacles too. Gauge your turns so you won't run up over curbs.

brakes

The braking system is one of the most important systems on your vehicle. Your life may depend on being able to brake effectively in an emergency situation. That means having brakes powerful enough to stop the vehicle you're driving, brakes that are compatible with each other and in proper adjustment. A surprising number of vehicles have faulty brakes.

Ensuring a properly functioning brake system begins when a vehicle's specifications are set and is one of the most critical maintenance tasks. Brakes out of adjustment are a contributing factor to many truck accidents.

What can you do to help prevent brake problems? One element you can check yourself is the slack adjusters. Some vehicles are equipped with automatic slack adjusters, while others must be adjusted manually. Whichever type you have, be sure they're functioning properly.

Especially before driving in hilly country, it's important to check the slack adjustment on your rig. All brakes should be adjusted to approximately the same degree, so that they are all doing an equal amount of work.

Good braking habits affect the life expectancy of the brakes on your vehicle. Don't "fan" brakes continually. This can reduce available air to an unsafe point and build heat. When you need to slow down on hills or in heavy traffic, gear down to help reduce your speed.

Anticipate traffic flow and the actions of other drivers so you can avoid braking any more than necessary. Try to drive with the flow of traffic.

How do you make a good effective stop? Firmly apply the brakes to begin with, then start easing up on them as you slow down, so very little pressure remains by the time you've fully stopped. One secret in being able to make good safe stops is to give yourself plenty of following room.

The old rule of allowing at least one vehicle length for every 10 miles per hour of speed is a good starting point. Given the limitations of truck braking systems and the amount of weight you have to stop, it's probably a good idea to give yourself even more room than the formula above would indicate. The more space you have in front of you, the more likely you'll be able to respond effectively if a hazardous situation occurs.

clutch

Your clutch compensates for the difference in rotation between the running engine and a stopped vehicle. In mathematical terms, when you engage the clutch you're putting 35 tons or more of weight into motion. Obviously, your clutch has a heavy-duty job to perform. Proper specifications and good maintenance are important in keeping the clutch operating well.

Your driving habits can do a lot to extend clutch life. If you master engaging the clutch properly, starting in the correct gear and not riding the clutch or slipping it, you'll add significant time to the life span of your clutch. Never engage the clutch while coasting; coasting in gear with the clutch released shortens clutch life too.

Whether your vehicle is empty or carrying a load, select the gear to start that lets you get the vehicle moving when the engine is at an idle. If you have to race the engine to get yourself moving, you're overtaxing the clutch unnecessarily.

The governor controls idle. When the clutch is released, the governor automatically adds fuel to maintain idle speed. There's no need to "give it throttle" to start. One method is to keep one foot on the brake while releasing the clutch until it just begins to engage. Release both pedals together. This keeps the vehicle from rolling back until it is under its own power. This method not only works well for starting on hills but also on flat terrain.

Matching truck speed and engine speed is the key to smooth shifting. If you don't watch the speed at which you change gears, you're putting unnecessary strain on the clutch and

experience speaking

"In this industry, where we have a lot of stop and go, our primary problem, in getting good driver habits explained, and in use, concerns riding the clutch. They have a tendency to leave the foot on the clutch in stop and go traffic in order to keep the vehicle in a moving position. They don't know that when they're doing that they're doing serious damage. And very costly damage. Very much reducing the life of the clutch."

John
Safety Manager
Tennessee

possible the engine.

Keep your foot away from the clutch unless you're actually in the process of shifting. Once you've completed a gear change, get your foot off the pedal fast. The minute there's any weight on the clutch pedal, the free play is taken up, and the bearing begins to spin, causing heat and unnecessary wear.

You need to be aware of the weight rating of your vehicle's power train. If it calls for 30,000 pounds, but loaded up to 50,000 pounds, you're asking the clutch to move an incredible amount of weight for which it simply wasn't intended. Overloading will cause premature clutch facing wear and early failure.

Your powers of observation can be especially helpful to your maintenance department in diagnosing clutch problems requiring adjustment. Unusual noise could be a signal that you've been riding or slipping the clutch, or it could just mean the clutch needs lubrication.

If the clutch slips, facings could be worn or there may not be enough clutch pedal free play. If there's drag, adjustment may be needed or (more seriously) a disc might be warped. Be as specific as possible in describing the problem to maintenance.

If the truck "bounces" when the clutch is engaged, it means there are hot spots or the clutch has warped due to excessive slipping. If the clutch pdeal falls to the floor, check the linkage.

Progressive shifting can reduce wear, save fuel and cut down on noise. What is progressive shifting? In ordinary shifting procedure, you accelerate all the way up to the governor before going into the next gear. In progressive shifting, however, you accelerate only enough to allow you to shift into the next gear, not up to the governor. You would bring engine r.p.m. up to the governor only if you required maximum power in a situation.

Good judgment is required to decide the appropriate point at which you'll shift gears. The necessary r.p.m. level will vary depending on whether you're shifting on the level, on an upgrade or on a downgrade. Shoot for a smooth and clean change of gears wherever you make the shift.

engine

Progressive shifting is just one example of several ways to maximize engine performance and prolong engine life. Treat your engine with care on the road, and it'll give you good service and troublefree operation.

Even with good driving habits, engines sometimes do develop problems. Probably the most obvious indicator of engine trouble is oil consumption. If your oil consumption goes to less than 200 miles per quart, you've got some kind of problem. If you're lucky, the problem will be repairable. Increased oil consumption is often accompanied by loss of power, increased fuel consumption and reduced compression.

Another indicator of trouble is injectors that become erratic; plugging, poor spray patterns and overfueling are usually the culprits. If you give the engine the throttle, it will act like it's choking or "missing".

Approach your time behind the wheel in a professional manner and realize the importance of your role in the fleet's preventive maintenance effort are the first two steps in making a contribution to preventive maintenance. Being sure your on-the-road techniques make the most of your vehicle without putting unnecessary stress on it is the rest.

preventive maintenance **review**

1. What is the basic principle of preventive maintenance?

2. What part can your powers of observation play in a preventive maintenance program?

3. How can you use your senses to monitor your vehicle's performance?

4. What part do your driving habits play in a preventive maintenance program?

5. What is progressive shifting?

6. How does a preventive maintenance program save a company money?

Trip Planning 12

the driver's role in trip planning

estimating time, fuel & expenses

map reading

five-step trip planning

Trip Planning 12

The objective of trip planning is to design the safest, most time and fuel efficient route between two or more points.

The penalties for poor trip planning are numerous, but they all come down to one thing, safety. Poor trip planning costs dollars and accidents.

Route selection should take several factors into account, including: type of trip, federal, state and local regulations; type of cargo; weather; traffic; and road conditions. You must pick the most direct route based on all of these factors. You should know before you set out that your load can be delivered on schedule with maximum economy. And without breaking the law!

Trucking companies wouldn't be able to stay in business if drivers and management didn't work together to plan the safest and most efficient routes in order to meet delivery schedules. Careful planning is required to reduce fuel usage, keep wear on the vehicle and all components at a minimum, and hold down maintenance costs. Running empty (deadhead mileage) must be limited to whatever degree possible.

Regulations requiring permits and licenses (used to raise tax monies for the building and maintenance of roadways) need to be taken into account during trip planning. So, too, do regulations restricting vehicle size and weight, type of cargo, night and day traffic, holidays and city truck routes.

The Driver's Role In Trip Planning

The driver's role in trip planning will vary from company to company. A driver working for a regular route carrier will most likely be travelling established routes. Deviating from that route may be strictly forbidden, with exceptions severely limited. Even driving for a regular route carrier, however, you should still be able to discuss trip planning with the dispatcher.

On the other hand, driving for an irregular route carrier is likely to require in-depth knowledge of trip planning. In some

experience speaking

"Once I've found out what I'm supposed to be doing for the trip, generally I will try to plan the length of time that it takes to make the trip. In a lot of cases, if I don't know where I am going, I'll call ahead and get a directions and also make sure that my receiving appointment is right. I check the weather conditions, if it's winter time, and pretty much plan out where I'm going to need to stop as far as fuel stops and my longer stops to sleep and so on. It makes the whole project an awful lot easier for me."

George
Driver
Wisconsin

cases, drivers plan their own routes. But even if the dispatcher lays out your route for you, you may need to come up with alternate routes in the event of an emergency. Getting assistance or advice from the dispatcher or other drivers can be a big help.

In planning a trip, consideration must be given to whether the trip is local or over the road, the length of the trip, and if the trip involves interstate or intrastate hauling.

driver factors

Primary consideration must be given, during trip planning, to your own abilities and limitations. You need to be properly rested before heading out. You have to be given a reasonable amount of time to make the trip in. Each trip has to be planned in such a way that you're able to comply with hours of service regulations. You have to plan for fuel, food and rest stops. And you need to develop reasonable expense estimates, so you don't end up stranded and out of cash.

vehicle factors

Number of trailers (singles, doubles, triples), number of axles required for gross weight, overall weight and length restrictions need to be considered. Clearance requirements or restrictions mean the height and width of your vehicle could be a factor in trip planning. Steep grades and curves may need to be avoided because of vehicle size.

You may need to scale your truck to be sure weight is distributed properly.

cargo factors

The type of cargo being hauled, weight and center of gravity all affect trip planning. Numerous restrictions might apply to hazardous materials loads. Liquid, top heavy and generally very heavy loads mean avoiding steep grades and curves. Traffic laws or local regulations relating to cargo may keep you from taking certain routes.

road conditions

Weather must always be taken into account when planning a

trip. Traffic conditions are equally important. Try to time your food and rest stops for during rush hours in heavy traffic areas. In this way, delays due to heavy traffic may be avoided.

Avoid low clearances (low underpasses and tunnels). Avoid narrow roads. Avoid older roads or bridges that won't support heavy vehicles. And, as much as possible, avoid steep grades and sharp curves.

Toll roads and nontoll roads need to be considered in terms of time saved versus expense (often the toll road is the most economical).

Avoid local traffic congestion by steering clear of shopping centers, schools and seats of government in larger metropolitan areas. Use limited access highways, which have an overall better safety record.

Route selection and planning depend on the type of trip, cargo and many variables. Failure to do a good job of trip planning can result in delays, economic delays or even an unsafe trip.

Estimating Time, Fuel & Expenses

Trip planning requires the development of practical skills. Estimating the time a trip will take is necessary for planning stops, knowing approximate arrival times and, most important, meeting schedules. Fuel usage estimates are needed to plan fuel stops and to make sure you carry enough cash or credit to cover fuel expenses. Estimating the amount of money a trip will take also means planning for tolls, food stops and overnight rest stops.

estimating trip time

There is one simple and generally reliable method for estimating trip time that's used by many companies and drivers. Allow 2. 5 hours for every 100 miles you will travel. Experience suggests this is reasonable for driving time, meals, fuel and rest stops.

$$\frac{\textbf{Distance x 2.5}}{\textbf{100 miles}} = \textbf{Hours}$$

Investigate weather, traffic and other conditions which affect average miles per hour. Keep safety in mind at all times. And avoid fatigue.

experience speaking

"Trip planning is very important. It saves you time. You regulate your time, and the first thing you have to know is the mileage of your trip. If it's an unfamiliar trip, after you have the mileage, get your atlas out. See what route you're going to take. Try to plan ahead where you're going to stop. If you are going on an 800-mile trip, you'd naturally have to take an eight hour break somewhere in that trip. So try to figure your hours out. Look at your delivery time. If you have a 7:00 AM delivery on the following day, back up the time you need for your trip. Your fuel: If you're carrying 200 gallons of fuel, you're going to be able to travel your 800 mile on 200 gallons of fuel. But, you're going to have to figure out your eat stops in there. Your first rest period or stop should be within the first 100 miles. You want to check your unit. You want to check your load that first 100 miles. If you had something you missed in your pre-trip, maybe a little squeal or a bad tire, whatever, you are probably going to pick it up on that first 100 miles. After that, plan how often you want to stop. And you don't want to stop too often. You want to plan your trip so that you make as few stops as you have to, because each stop you make takes an awful lot of time out of your trip."

Mel
Driver
Nevada

experience speaking

"Trip planning is important, because it enables you to determine the time factor on a pickup or delivery. It enables you to plan timing through cities during business hours. It gets you a chance to stay on schedule to make your pickup and delivery on time. Which is important to the people you work for."

Frank
Instructor
Louisiana

No driver will consistently be able to average the maximum speed limit without breaking the speed limit. Stretches of road will be slower due to weather, hills, traffic and other factors.

estimating fuel usage

The first step in estimating fuel usage is to determine the range of the vehicle. To find the range of your vehicle, multiply the tank capacity in gallons by the miles per gallon that vehicle delivers.

Tank Capacity (gallons) x mpg = Range

Average fuel usage is based on highway driving time, intown driving time, unexpected slowdowns and idling time (at zero mpg).

Fuel use estimates are important to plan fuel stops efficiently and as one factor in establishing a trip budget.

estimating expenses

A driver must plan ahead for the money required to take a trip (expenses). You must consider the distance to be travelled and the time the trip will take, in addition to the potential need for emergency funds. The most obvious expenses are meals, layovers, rest stops, fuel stops and tolls. If the weather is bad or bad weather is forecasted, you should carry enough cash or credit to cover tows and unexpected layovers. The latter two are things no driver wants to deal with, but in severe weather conditions they're often part of the standard cost of doing business.

Time, fuel and money estimates are a key part of trip planning. The driver needs to investigate procedures depending on the overall type of driving situation involved.

Map Reading

Professional drivers often need to use maps to determine routes or locate specific destinations. The obvious consequences of not using maps as a travel aid are getting lost, losing time and making late deliveries. The best drivers always know exactly where they're going and all the details involved in getting there.

Essential map reading skills for the professional driver include the following:

Locate starting point, intermediate stops and destination;

Lay out route on a map;

Estimate point to point mileage using the map scale;

Read map symbols.

key characteristics of maps

The face of most maps is set on a box, or grid, pattern. Letters are spaced at regular intervals along top and bottom. Numbers are placed along the sides.

Scales vary from map to map. The scale for each map is explained on the map. Each scale gives a number of inches to represent a given number of miles. While map scales can be used for rough estimates, mileage charts (when available) give a better idea of true distances. The most skilled drivers can and do use both.

Symbols vary from map to map. Highway maps are more general. City or regional maps often include parks, public areas, schools, churches and other landmarks. All maps carry a legend explaining the symbols used.

sources of maps

Local governments, chambers of commerce and tourist organizations often provide free maps. Commercial maps and atlases are available from many distributors, including most truck stops.

types of maps

Highway maps show entire states or regions. Some major cities are usually shown in some degree of detail. Features vary depending on the map, but you can expect to find major routes, road types, toll roads and freeways, population centers, railroad crossings, junctions, and the mileage between junctions. Highway maps are best used to locate major highways and routes for long distance travel.

City maps provide the greatest detail. Major street names are readable. Although numbers may be very small, street ranges are usually indicated to help pinpoint exact locations. City maps can be very helpful in plotting the last leg of a trip.

experience speaking

"You have to know the things in between, in other words, such as setting up your routes. Where your scales are at, where your port of entries are at. You have to regulate it to understand that the west coast and the east coast, for example, do not receive on weekends. So consequently, you've got to regulate and pre-plan your trip so that you end up there and are able to make your delivery instead of laying over until Monday. It's really pre-planning the route, planning your stops, knowing your company's fuel stops. At the same time, of course, when you're trip planning you want to make dog-gone sure that the routes you're taking are routes that you can handle with the rig you're operating. Check your height, your weight, what have you. Find out if the states you're going through have overpass or bridge restrictions. Always pre-plan, and you'll have a safe trip and a fast one."

Robert
Instructor
Missouri

Many larger cities are also covered by city, county or region atlases. The detail is often excellent. One disadvantage with such an atlas, or book, is that it may take several maps to route a trip through one city.

using a map

One excellent way to plot a trip is to make your own map. This can help you avoid making stops for map checking.

After studying the appropriate maps, draw your route on a single sheet of paper using as many main roads and streets as possible. Put in towns between, your standpoint and destination along with highways, road and street numbers and names, as well as point to point mileages. It's still a good idea to carry other maps for reference purposes, but the better you get at making your own maps, the less you'll need other references. To make reading easier (and keep your eyes on the road), print in large letters.

Five-Step Trip Planning

There are five basic steps involved in trip planning, no matter what type of trip is involved.

Step 1: Make sure paperwork is up to date.

To meet regulations, obtain permits, licenses and cab cards (company officials' and/or driver responsibility).

Driver logs current (solely driver responsibility).

Freight documentation (driver responsibility once on the road).

Step 2: Select the route (Dispatcher and/or driver responsibility).

Many variables involved depend on situation.

Vehicle and cargo restrictions (for example, length, width, clearance, type of cargo).

Hazardous materials restrictions.

Best route for type of cargo.

Local traffic conditions and regulations.

Weather and other environmental conditions.

Step 3: Estimate time and plan for rest stops (company and

driver responsibility).

Driver safety is important.

Need to plan trip to meet schedule, but rest stops have to be considered.

Step 4: Estimate fuel, and plan for fuel stops (company and driver responsibility).

Step 5: Plan money and other personal needs (company and driver responsibility).

Must understand nature and length of trip.

Must know conditions to expect on the road.

Must understand company policy about reimbursements.

Must plan for personal items (money, clothing, safety and maintenance tools).

trip planning review

1. What is the objective of trip planning?

2. Name some of the penalties of poor trip planning.

3. What factors must be taken into account during route planning?

4. What is the driver's role in trip planning?

5. Estimating time, fuel and expenses is the responsibility of the dispatcher: True or False?

6. What is the formula for estimating trip time?

7. Identify four essential map reading skills.

8. The best way to plot a trip is to make your own map: True or False?

Hours of Service 13

hours of service regulations

10 hours driving time

15 hours on-duty time

60 or 70 hours of service maximum

recording your time

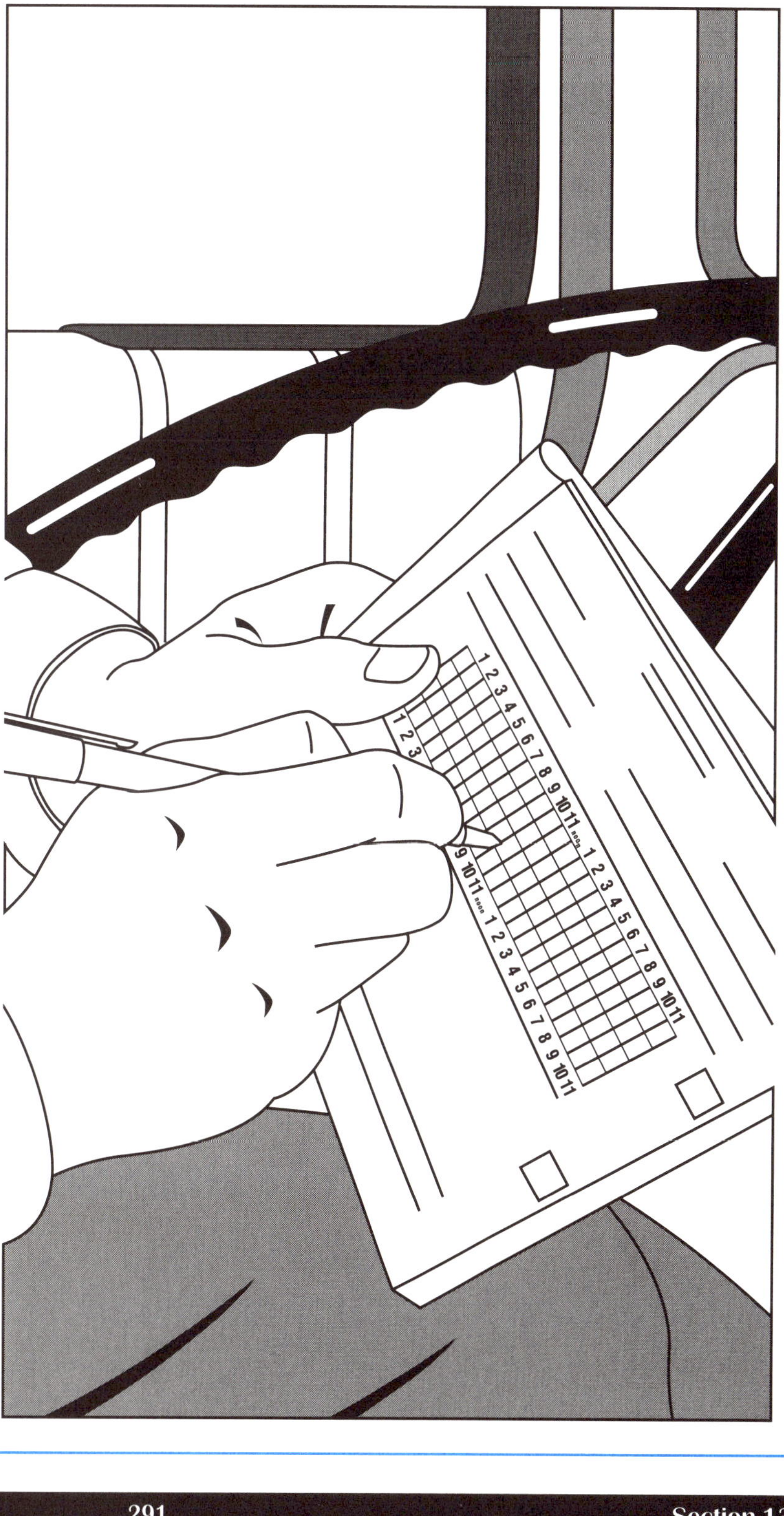

Hours of Service 13

A tired driver is not a safe driver. That's why the hours of service regulations exist.

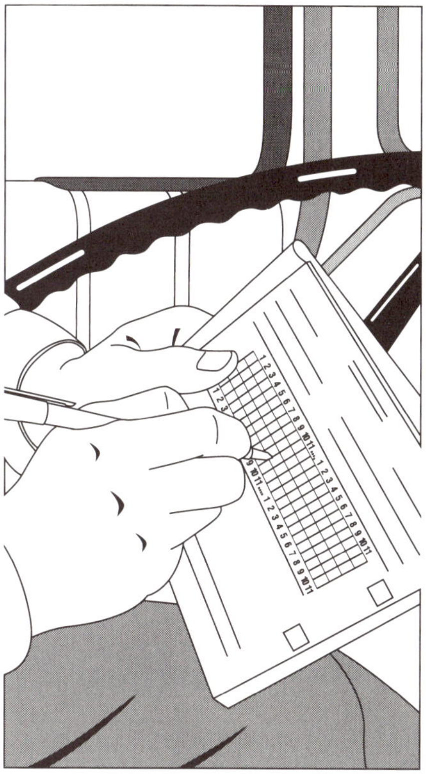

Mandating limits on the time you can spend behind the wheel before you're required to have a period of time off makes sense. This prevents fatigued drivers from operating vehicles. The hours of service regulations are designed for the protection of both the driver and the public.

There is no specific required form on which a driver's time must be recorded, only the requirement that certain items of information be recorded in some fashion. The most common way for drivers to record their time is in a logbook, officially known as the record of duty status.

Hours of Service Regulations

On the federal level, hours of service of drivers are part of the Federal Motor Carrier Safety Regulations (FMCSR), issued and enforced by the Office of Motor Carrier Standards in the Federal Highway Administration of the DOT.

Specifically, Part 395 of the FMCSR covers Hours of Service of Drivers and gives the requirements and exceptions that apply. You must become familiar with its provisions. In addition, many of the individual states have hours of service requirements that apply to operations within their jurisdictions.

These requirements may match the federal requirements in whole or in part.

Hours of service regulations apply to all carriers (and their drivers) involved in the transportation of property or passengers in interstate or foreign commerce.

The only major exception listed in the regulations is vehicles which have a manufacturer's gross vehicle weight rating of 10,000 pounds or less, unless they are hauling hazardous materials which require placarding. Drivers should read the regulations in Part 395 of the FMCSR to become familiar with all the details.

federal requirements

The FMCSR regulations on hours of service concentrate on three basic limits: 10 hours of driving time; 15 hours on-duty time; 60 or 70 hours of service maximum (depending on the type of operation you work for).

10 Hours Driving Time

The basic rule is very clear. Behind the wheel time is driving time. After 10 hours of it, you must have eight hours of rest. In order to have 10 full hours of driving time available to you again, the eight hours off-duty must be in a consecutive block of time.

sleeper berth option

A useful option is available that is especially helpful for team drivers on a long run. Done correctly, you can make use of the sleeper berth to accumulate off-duty time so you and your partner can keep your vehicle on the road for the maximum amount of time without a layover.

The following is for purpose of example only, and describes an ideal situation. Your real experience will not be exactly the same, but the following situation shows the principle to use in making the most of your available driving time.

Theoretically, you could drive for five hours, then spend four hours in the sleeper berth, then drive another five hours, then spend another four hours in the sleeper berth. In this hypothetical case, you're satisfying the off-duty requirement.

The only thing you have to be aware of if you use the sleeper berth option for accumulating off-duty time is that a period in the sleeper berth must be at least two hours long, and there can be only two of them used to accumulate the eight hours of required off-duty time. You can use the sleeper berth for periods of less than two hours, but those periods don't count if you're trying to get your eight hours off-duty so you can drive.

After the last of your two off-duty periods (that amount to a total of eight hours) you do have driving time available again. But you don't have a full 10 hours of driving time available to you at this point. You have to subtract your last period of driving time (the time between your two periods in the sleeper berth – in the case of our example, five hours) from 10 hours.

That means you have five hours of driving time remaining.

This subtraction rule holds true until you have a complete block of eight hours off-duty at one time.

Then you have 10 full hours of driving time available to you once again, as long as you're still within the overall on-duty time boundaries.

The key is to keep close track of where you are in hours remaining to drive and number of hours required to accumulate necessary time in the sleeper berth. Determine how you can manage to have the shortest amount of downtime when your rig can't be on the road because you are not in compliance with the hours of service requirements.

bad weather exception

If you encounter severe weather, bad roads or traffic conditions that cause you to be delayed and that could not be foreseen when you set out, the 10-hour driving time limit can be extended an additional two hours. This should be regarded as a very exceptional case. It's a way to avoid the problem of hours of service violations in certain specific situations.

The exception doesn't apply if you were aware of the potential for hazardous driving conditions before you set out. You get up to two extra hours of driving time if, and only if, you could have made the run in the 10-hour limit under normal conditions.

15 Hours On-Duty Time

You can no longer drive after you have been on-duty for 15 hours until you have eight consecutive hours of off-duty time. The same safety principle is behind this rule as the driving time rule: A tired driver is not a safe driver. After 15 hours on the job, you are quite likely to be tired and less alert.

This rule does not preclude drivers working in other capacities for longer than 15 hours. You just can't drive after 15 hours on the job. The rule is often misunderstood to mean that a driver can only be on-duty for 15 hours, period.

Besides your actual time behind the wheel, you are on-duty whenever you are either working or required to be in readiness to work. All on-vehicle time (except when you're in the sleeper berth) is considered on-duty time, even when you aren't driving.

In addition, time spent doing your required inspections or troubleshooting vehicle problems, loading, unloading or making deliveries, attending a disabled vehicle or carrying out your duties at an accident scene are on-duty time. Even work that has nothing to do with transportation is included in on-duty time if it's required by your company. Work at other jobs you might hold part-time must also be counted, even if driving is not involved.

60 or 70 Hours of Service Maximum

The third hours of service rule varies according to the length of the work week used for recording time by the trucking operation you work for. Generally speaking, companies either operate with a six-day week or a seven-day week.

If your company uses the six-day week, you're allowed to accumulate a maximum of 60 hours on-duty time in any seven consecutive days. If you work with a seven-day week, you can accumulate 70 hours on-duty time in any eight consecutive days.

Recording Your Time

There has been no prescribed log form since January 1, 1983. You are only required to maintain a record of duty status that accounts for all your time and includes certain specific items and a grid for recording time.

time recording options

You may find your company using one of several options for time recording.

A driver's log; standard form or modified.

Driver's multi-day log, standard form or modified.

Vertical graph grid combined with a company record.

Horizontal graph grid combined with a company record.

Most companies require drivers to keep a daily log.

11 required items

Regardless of the form your company prescribes, it must include the following 11 items of information.

The graph grid with a remarks section (usually located below or next to the horizontal or vertical grid).

The date.

Total miles driving today.

Truck or tractor and trailer numbers.

Name of the carrier.

Carrier's main office address.

A place for driver signature/certification.

The starting time of the 24-hour period covered by the log. This starting point can be set anywhere at the discretion of your company, but the same starting point must be used for an entire terminal.

Name of co-driver (if there is one).

Total hours (at end of grid).

Shipping document number or name of shipper and commodity.

A record of duty status must be filled out in duplicate for each 24-hour period. Entries relating to your duty status must be legible and in your own handwriting. Some of the other information required on the record may be preprinted (carrier name and main office address, for example).

100-mile exemption

If you're a driver who operates within a 100 air-mile radius of your normal work reporting location (and you aren't a driver salesperson), you may be exempt from keeping duty status records as long as your company keeps very careful records (remember that you're still subject to the other hours of service regulations). The regulations specify a number of conditions that must be met for the exemption to be valid. Check with your supervisor.

Another element of the 100-mile exemption is that you must be released after a 12-hour work day. If you are required to be on-duty longer than 12 hours, you'll be required to log, even if you operate within the 100 air-mile radius.

lightweight vehicle exception

If you're a driver of a lightweight vehicle (gross weight of 10,000 pounds or less), you're also exempt from keeping a record of duty status unless you're hauling passengers or hazardous materials in a quantity that requires placarding.

retention of records

You're required to keep the duplicate copies of the records of duty status for the previous seven days in your possession. The copies must be available whenever you're on-duty. So at any time you must have your current record and the seven previous ones.

The original (which may be handed in or mailed to the terminal or main office, depending on company policy) must be retained for six months. You're required to submit records of duty status within 13 days of the period for which they apply.

keeping records current

The safety regulations require that a record of duty status be "current to the time shown for the last change of duty status." That means keeping your log or other format up-to-date at all times.

It's to your advantage to keep close track of your time so you know precisely how much driving or on-duty time you have remaining before off-duty time is required.

vacations (and other time off)

If you're off-duty for a whole weekend or go on vacation for one or more weeks, you don't have to fill out records of duty status for each individual day during that time off. But you do have to account for that time in some way for your recordkeeping to be absolutely correct. Follow company policy.

The method most often used is to fill out a single log or other record that shows off-duty time for the entire period of time indicated on the single form. This is sufficient record of your off-duty time.

working for multiple carriers

If you drive for more than one company, you're still bound by the hours of service regulations. You must follow the driving and on-duty time rules.

Make a copy of your record of duty status for each carrier that employs you. Show exact times worked for each company. Tell your supervisor how many hours you've been on-duty during the past seven days and what time you went off-duty at your last job.

driver's responsibilities

If you're an interstate driver transporting property or passengers, you must comply with the hours of service regulations. Your company is required to provide the format on which to record your time, but you must fill out the specified format and keep it accurate and up-to-date.

The federal regulations place responsibility for accurate time recording on you as well as your employer. Federal motor carrier enforcement personnel can issue citations for hours of service violations, and fines can be levied on both carriers and drivers. You can also be placed out of service for logging violations.

Know the requirements and follow them. Ignorance of the rules and regulations is no excuse for non-compliance. You can be cited just the same as if you're willfully violating the rules.

Know the hours of service regulations. Then make it a practice to fill out your daily log carefully, neatly and correctly. Keep it legal and keep it current.

hours of service review

1. What is the reason for hours of service regulations?
It prevents tired drivers from operating vehicles which protects both the drivers & public.

2. Hours of service are part of the FMCSR: True or False?

3. After 10 hours of behind the wheel time, how many hours must be spent off-duty before driving again?
8

4. What is the maximum on-duty time in any given day?
15 hrs.

5. The sleeper berth may be used to accumulate off-duty time: True or False?

6. What is the most common method used for time recording?
a logbook

Drugs & Alcohol 14

the laws

Drugs & Alcohol 14

Smashed, bombed, stoned, plowed, buzzed, ripped. These are all words used to describe drug and alcohol highs.

Unemployed, broke, homeless, alone. These are all words used to describe the results of drug and alcohol abuse.

Even if you steer entirely clear of alcohol and drugs, you're involved in the issue.

Each time a news report shows that a truck driver was drinking or doing drugs when involved in an accident, it reflects on everyone involved in the profession. You owe it to your profession to help keep trucking's reputation clean. You may think of it as an issue of fairness or self-protection. Statistics show that: Drug users are three to four times more likely to be injured on the job; they're 2-1/2 times more likely to miss work; and they're more likely to steal from the business and co-workers.

Much of the public's attention has been focused on illegal drugs like speed, crack, cocaine, and marijuana. Aside from any medical effects, a drug is a chemical substance that changes your mood, perception, reaction or state of consciousness.

Most drugs act on the central nervous system, either speeding up (stimulants) or slowing down (depressants) mental and physical functions. Hallucinogens are a family of drugs that alter perceptions.

You must also be cautious with prescription or over-the-counter drugs.

Under federal regulations, you're subject to random selection for testing under your company's random drug testing procedures at any time, unless your vehicle is under 26,001 pounds or you're an intrastate driver in certain states. Chances are that you'll be tested at some point, anyway, either under that program or some other.

The Laws

Even if social or moral reasons don't sway your opinion, alcohol and drug abuse is prohibited by the Federal Motor Carrier Safety Regulations (FMCSR). A pre-existing condition of alcoholism or drug abuse can keep you from getting a license or a job. Succumbing to alcohol or drug use can get you disqualified from driving, perhaps for life.

The basics are laid out in the FMCSR, which you should read. The FMCSR states that: disqualifying offenses include:

> Driving a commercial motor vehicle while the person's alcohol concentration is 0.04 percent or more; driving under the influence of alcohol, as prescribed by State law; or refusal to undergo such testing as is required by any State or jurisdiction.

The enforcement level on a driver's alcohol concentration, sometimes called Blood Alcohol Content (BAC), varies among the states with many states still considering a person legally drunk if his BAC is 0.10 percent.

The federal standard of 0.04 percent is much more strict than most states. However, that's not the point. Tests show that your ability to function competently begins to be impaired long before the 0.10 level. Many people are severely limited in abilities with a BAC of 0.05 percent.

The FMCSR not only prohibit on-the-job use of alcohol, but also limits your alcohol consumption by requiring that at least four hours pass before you go on duty or drive after you consume an alcoholic beverage. Part of the burden also falls on your employer by requiring that they make sure you observe the rules.

Additional causes for disqualification include driving a motor vehicle under the influence and transportation, possession or unlawful use of any illegal drug (included in the Schedule 1 list, Table of Disqualifying Drugs and Other Substances).

Disqualification from driving can range from six months to life.

drug testing

If you've been driving for any length of time, you already know about the drug testing regulations, if they target you or not.

If you're a new employee with a motor carrier, you were tested before you were hired. Existing employees were tested at the time of their next scheduled physical exam. You may already have been selected under the random testing procedures. Drug testing is also required after DOT reportable accidents in which you receive a citation. The regulations provide your employer with the opportunity and duty to have you tested if there is a reasonable cause to suspect drug abuse.

As this manual is being written, the Federal DOT was considering similar regulations for testing of alcohol usage as well.

The federal regulations require that you, your supervisors and company officials be trained in drug abuse issues. They also include provisions for a motor carrier Employee Assistance Program (EAP) to be established.

employee assistance programs (EAP)

The drug testing regulations specify some kind of EAP, but they stop short of mandating a complete employee assistance program. If you are fortunate, the company you work for will have an employee assistance program in place.

EAPs are usually designed to help employees solve problems and can provide motor carriers with a method for dealing with driver problems, such as drugs.

A good EAP serves as a gateway to the wider resources of your community. EAPs are staffed with psychologists, social workers, or other specialists who can assess the needs of an employee and refer that person to an effective source of help. That help could include counseling, treatment or rehabilitation services in the case of substance abuse.

Between the time the federal government issued its first proposals and the final rule in 1988, it determined that motor carriers will not be required to offer an opportunity for rehabilitation or to provide job security to drivers who fail a drug test, use drugs on the job or who voluntarily request rehabilitation.

The federal DOT decided that issues such as firing, hiring temporary drivers, reassignment and other issues and policies were more appropriate left up to the private sector. So, these issues are not covered in the federal regulations.

drugs and alcohol review

1. Drug users are three to four times more likely to be injured on the job: True or False?

2. You can lose your license and be disqualified from driving for life for drug abuse: True or False?

3. What are some specific causes for drug or alcohol related disqualification?

4. Are motor carriers required to offer rehabilitation programs to convicted drug or alcohol abusers?

5. What is an EAP?

6. At what point, during use of alcohol, does your ability to function competently begin to be impaired?

Driver Qualification 15

basic driving qualifications

physical qualifications

driver disqualification

Driver Qualification 15

Prior to receiving your Commercial Driver's License, you should read and become familiar with the Federal Motor Carrier Safety Regulations (FMCSR).

This is an important stepping stone on the path to becoming a professional driver. The exact wording of the regulations concerning driver qualification can be found in Parts 383 and 391 of the FMCSR. The information that follows here is a summary only.

Federal regulations require the use of qualified drivers on property and passenger carrying vehicles in interstate commerce, with only a few exceptions.

Basic Driving Qualifications

There is a basic set of criteria which must be met by any driver who is subject to the FMCSR. The safety regulations specify that you are qualified to drive if you meet all of the Federal requirements listed here.

You are at least 21 years old.

You can speak, read and write English well enough to do your job.

You can drive your truck safely. You can tell if your truck is safely loaded.

You know how to tie down your cargo.

You can pass the DOT physical exam, including the drug testing requirements.

You have only one up-to-date driver's license.

You have taken a DOT written exam on the drivers' regulations.

You have passed a truck driver's road test.

You have filled out a detailed job application.

You have a good traffic record.

If you drive a vehicle that weighs 26,001 pounds or more

(GVWR), carries 15 or more passengers or transports hazardous materials, you must have a CDL from your state which meets Federal standards.

Any time you're convicted of a violation, you must notify your employer and the state that issued your license (if the violation was committed in another state) within 30 days.

Physical Qualifications

A major element of driver qualification is the physical examination. Since your physical and mental condition plays a big part in your safety on the road, this exam is extremely important. If there's any one aspect of driver qualification that's stressed more than others, it's the physical exams; the Federal regulations specify the basic content of the physical examination.

Physical handicaps (loss of a limb, for example) are a big concern. Certain diseases and medical conditions requiring medication or that may cause unexpected problems are also looked at carefully. Finally, drug and alcohol problems can be a factor in the physical qualification process.

As with the road test and written exam, your potential employer has an option when it comes to the physical exam. He can either require a physical that meets the standards or he can accept a currently valid Medical Examiner's Certificate, if you have one in your possession that was issued within the previous 24-month period.

If you pass your physical exam, and you don't already have a Medical Examiner's Certificate, the doctor will give you one. You must carry the Medical Examiner's Certificate with you at all times.

eyesight

You must have good eyesight. You must be able to see far away objects. You must also have good side vision. You can wear glasses or contacts that correct the way you see. You must be able to tell traffic signal colors apart. You will not be qualified to drive if you have only one eye.

hearing

With or without a hearing aid, you have to be able to hear a loud whisper at five feet with your better ear. Or, you have to pass the hearing test on a doctor's testing equipment.

drug use

Drug use will keep you from being qualified. You can't use amphetamines, narcotics or other dangerous substances or habit-forming drugs. Also, you can't use drugs which you regularly (and legally) use that may keep you from driving safely. Finally, you can't be an alcoholic.

good health

Some diseases will keep you from driving safely. You won't be allowed to drive if you have any of the following diseases or disorders.

Diabetes, if you need insulin to control it.

Heart disease which causes you chest pain, fainting or shortness of breath.

Chest or breathing problems, such as asthma, emphysema and chronic bronchitis.

High blood pressure problems.

Loss of movement or feeling in part of your body.

Any sickness that may cause you to faint.

Mental problems or severe nerve problems.

getting a waiver

If you have lost, or can't use a foot, leg, fingers, hand or arm, or if you have a bone or muscle problem, you can't drive without a waiver. To apply for a waiver, you must, individually or jointly with a motor carrier, send a waiver application to the appropriate Regional Director of Motor Carrier Safety.

medical examiners certificate

The importance of physical qualification is underlined by the requirement that you must carry your Medical Examiner's Certificate of physical qualification on your person at all times while you're driving.

Once you're qualified and have a valid medical certificate, it remains in effect for a period of two years (24 months). After that time you must be re-examined by a qualified medical practitioner. The re-examination requirement is a logical one since your state of health can change considerably in a two-year period.

Driver Disqualification

Disqualification of a driver can come about as a result of information that surfaces in the various reports, reviews, and records required to be done by the motor carrier on a regular basis. Certain specific conditions must exist for a driver to be disqualified from driving a commercial vehicle.

Once a disqualification notice is given to a driver, he or she is prohibited from driving for the length of time specified in the notice. If this happens to you, and you're fortunate, your company will assign you to non-driving tasks for the duration of the disqualification. However, disqualification could also mean sitting it out for the duration.

cause for disqualification

Disqualification is required under terms of the safety regulations in the following situations.

Conviction or forfeiture of bond or collateral for the following:

> Operating a motor carrier vehicle under the influence of alcohol;
>
> Operating a motor carrier vehicle under the influence of drugs;
>
> Transportation, possession or unlawful use of drugs;
>
> Leaving the scene of an accident resulting in personal injury or death;

A felony involving the use of a motor vehicle.

Loss of driving privilege due to revocation, suspension, withdrawal or denial of an operator's license, permit or privilege for:

Speeding or reckless driving;

Motor vehicle law or ordinance violations;

Disregard for public safety.

Failure to pass the required physical examination and obtain a valid medical examiner's certificate.

Although the list of situations above is specifically cited in the safety regulations as requiring disqualification, your motor carrier may disqualify you for other reasons as well. An example might be excessive logging violations, or other situations where the employer feels his or her best interests or the interests of highway safety aren't being served.

If your carrier disqualifies you for any reason, the recommended procedure is for a written notice of disqualification to be prepared and delivered to you, with a copy going into your personnel file as well.

Printed forms are available for this purpose which specify the reason for the action, including a reference to the appropriate section in the regulations, the date and the carrier's intended action.

period of disqualification

In general, a first offender won't be allowed to drive for one year following the conviction. If you transport hazardous materials, you're disqualified for three years. If convicted of a felony involving the manufacture, distribution or dispensing of a controlled substance, you're disqualified for **life**.

A second conviction for driving under the influence, leaving the scene of an accident or committing a felony using a commercial motor vehicle disqualifies you for **life**.

traffic violations

Serious traffic violations involving commercial motor vehicles can also result in disqualification, even if they don't result in license revocation. For example, you can be disqualified for at least 60 days if you've committed two serious traffic violations within a three-year period. Three serious violations in the same period will result in 120 days of disqualification from driving.

The point of all this information is that you need to take the matter of driver qualification and disqualification seriously. The principle behind the system is safer roads for all of us. Know the rules and follow them religiously whenever you're on the job. Stay qualified and out of trouble.

driver qualification review

1. What are the FMCSR?

2. You must have a CDL to drive a vehicle that weighs 26,001 or more pounds (GVWR): True or False?

3. Will certain physical handicaps keep you from being qualified as a tractor-trailer operator?

4. What is a Medical Examiner's Certificate?

5. Name three causes for disqualification.

6. Can you be disqualified for traffic violations?

The Trucking Professional 16

trucking industry image

contact with the public

customer relations

basic job requirements

applying for a job

tests

The Trucking Professional 16

A great deal of attention is being paid to the importance of professional drivers these days.

Over and over again you hear motor carriers say that their drivers are the real key to the success of their operations. And without the professional driver, no preventive maintenance or safety program can work. It's the driver who makes the industry, as well as the company succeed. When customers or the general public see trucking as a safe and important part of our economy, the driver can be thanked. But it's the driver, too, who can create a tarnished image of the whole industry.

Trucking Industry Image

Truck drivers who are road hogs, speeders, tailgaters or who are generally rude in their behavior, in and out of their vehicles, are responsible if the public has a poor image of the industry overall. The consequences of a poor public image are many. Letter writing and phone campaigns, public and media outcry can have a strong effect. Putting your job in jeopardy becomes almost a minor issue compared to the result of restrictive legislation. Eventually, it becomes more difficult for trucking to be profitable, which inevitably results in the loss of jobs. Whether you want to hear it or not, one driver's negative performance can have widespread impact on the industry.

Fair or not, all laws can be influenced by the public's image of the trucking industry. The trucking industry is subject to many Federal and State laws and regulations. The vast majority of laws are designed to, in some way, protect the public. Some laws, such as road use taxes and operating restrictions, make trucking costly and often difficult. That isn't to say they're unfair. Many people talk about the days of the "knight of the road" who operated safely and always stopped to help motorists in trouble. Those days haven't disappeared completely, nor have the kind of drivers who created that image in the first place. The professional drivers who know and practice safe, courteous driving at all times are still out there. They're probably in the majority. They aren't as easily noticed or long remembered, however, as the few bad seeds we always hear about.

Contact With The Public

experience speaking

"We look for the person first. The philosophy of our company is hire the person first, the driver second. One thing is, you can never change the person. You can control or correct bad habits of drivers, but you can't change the person. So we look at the person first, the driver second. We feel that the driver is the last aspect of every sales effort our company makes. They are the ones that have the last contact with the customer when they make that delivery. If they make that delivery in a timely way, that's how the customer remembers us."

Dennis
Company Executive
Wisconsin

a good appearance

There are many areas of contact and conflict with the public in which the driver must be careful of his image and actions. The best place to begin presenting a positive image is to obey all laws. This alone will cut down on incidents that make life more difficult for you, your fellow drivers, and your industry. Next comes the appearance of the driver and his vehicle. A clean, neat appearance projects a good professional image. There's nothing wrong with looking however you want to look, but the fact is that people do, and probably always will, make judgments based on appearances. That's just the way it is. It also helps to stay in good shape physically and mentally. You look better when you're in good shape, in addition to being better equipped to handle the many physical and mental stresses placed on the professional driver daily. Your vehicle should be clean and well maintained too. Dirty vehicles, in need of repair, with flapping tarps and dragging chains say to the public that you don't care. Do your pre-trip and other inspections to ensure that your rig is in good shape. Keep it clean so others can see what good shape it's in.

sharing the road

Most conflicts with the public have to do with sharing the road. This is an area where it seems everyone likes to find fault with everyone else, and few take a reasonable, objective view. Since we all have no choice but to share the road with each other, it would be best if both sides simply always tried to do their best. There are a number of common causes for resentment toward truckers from the public point of view. Unfortunately, almost all of them are based on an experience. Other drivers feel threatened when large vehicles follow too closely or drive in packs that make passing difficult. Speeding trucks, cutting in and out of traffic, show a lack of concern for safety. The same is true of trucks that pass without signals or cut off other drivers.

The high beams of trucks are more annoying than those of cars. Dim your lights when meeting or following. Don't flash your brights to force a slower driver to move over. Take it slow and easy at intersections. Don't jump the light on left hand turns. Not only is this unsafe, but it can prevent the proper flow of other traffic. Wait your turn. On right hand turns, make sure your way is clear. These are some of the most common issues. There are others. Your best bet is to always operate in a safe, legal and courteous manner. We all know that car drivers also do many things that annoy truckers. Here again, it's your responsibility to operate safely, legally and courteously. Avoid the temptation to retaliate when another driver's actions bother you. Stay calm. Act like a professional. Your negative reactions could make a situation even more unsafe. Offensive actions could get you fired. The public has learned to identify and report unsafe or offensive truck drivers, and they often do.

meeting the public

Be courteous, polite and helpful. Avoid arguments, even if the other guy was in the wrong. Be a good representative of your company and your industry at all times. Helping others creates the good image the industry desires. Within the limits of company policy, offer as much assistance as possible. Stop and find out if stranded or broken down motorists need help. Give directions willingly to lost motorists. Essentially, treat other drivers as you would like your family to be treated. At an accident scene, give whatever assistance you're capable of.

projecting the image

When you have a chance, stand up for the trucking industry. State your case calmly, and back up your opinions with facts. Take every opportunity to tell others about your industry and what it takes to become a professional driver.

Customer Relations

As much as you are a public relations person for your industry in dealing with the general public, you are that and more to your company's customers. A customer's impression of you can lose or gain business for you and your company. Many companies see their drivers as driver-salesmen. It isn't hard to make the extra effort to do a good job and leave a good impression. Be prompt for pickups and deliveries. Be courteous. Don't argue or let yourself be provoked. Be honest and helpful. Always thank the customer for his business. Follow company rules in handling cargo and documentation. Know your company's procedures for dealing with freight problems, as well as dealing with customers.

experience speaking

"The first thing it takes is the right attitude. You have to be able to develop a positive attitude. You have to be able to pre-plan. It takes the proper paperwork. It takes you keeping your equipment safe and operational. It takes, above all, safety. Safety first, always. Sacrifice safety for nothing. You'll be on time. You'll make a living. You'll keep your equipment in shape, and you'll be able to drive tomorrow."

Frank
Instructor
Louisiana

Always be on the lookout for freight the customer is shipping that your company can handle. Know ahead of time if you can ask customers for their freight. More freight means more business, which in turn means a greater need for drivers and job security for you.

Basic Job Requirements

Some of the basic requirements were covered earlier in this manual. Those are based largely on federal regulations. Other requirements are laid out by employers or even placed on employers by federal and state regulations. Before being hired, you must fill out an application that meets DOT requirements. The prospective employer is required to conduct a background investigation for the previous three years of your driving record. As an applicant, you should know the various types of vehicles used by the trucking industry, with a basic understanding of systems and components. You should understand industry jargon (the language of the industry). Even though a company will teach you how to handle its paperwork, you need to know something about paperwork in general ahead of time. You must be able to drive vehicles assigned safely. And, lastly, you need a basic understanding of cargo handling techniques and procedures. If you plan to apply for a job as a truck driver, you'll need a positive attitude toward the trucking industry, the job and your prospective employer. A willingness to follow company policies and meet requirements is essential. You must have a personality that suits the job. That means having a high interest in the job, being mature and responsible, having a good safety attitude and being enthusiastic. When asked what the number one requirement for a good driver is, fleet managers consistently put attitude first. The idea is that you can teach someone to be a good driver, but a positive attitude is something you develop on your own. A strong, positive attitude will go a long ways in helping you find a job in trucking.

company policy

Each carrier that hires drivers has its own policies and ways of doing things. When you go to work for a carrier, find out what's expected and abide by it. Keep in mind that company policy takes precedence over procedures learned in a driver training program. This, of course, is with the exception that no one can force you to perform in an unsafe manner or break the law. Company policies will cover hours of work and benefits, work rules, safety rules, vehicle inspection and maintenance requirements, rules for trips and relations with customers and the public. The more you know in advance or find out during an initial interview, the better you will be able to evaluate a job against your own abilities and desires.

opportunities for advancement

As much as most newly trained drivers might like it, seldom will a company put an inexperienced new hire behind the wheel of a big rig. Over-the-road jobs that offer the highest pay also require the greatest skill. They come with experience. They have to be earned. The simple fact is that a motor carrier can't afford to put its most expensive equipment and its most valuable cargo in the hands of inexperienced drivers. Recognize going in that your education is really only beginning after you've gone through a truck driver training program and received your CDL. Often your first job with a company will be hauling local freight or even working in the yard, spotting trailers for loading and unloading. You might even start out as a freight handler with no driving duties. Whatever duties you're given, do a good job, show enthusiasm and learn all you can. Make the most of each step up. When opportunities come up, take them. Driving jobs with private carriers, in particular, require a lot of skill and experience. Often drivers will be called upon to act as driver-salesmen. Here again, attitude is a main consideration. These are tough jobs to get, but for many they're ideal. Most opportunities for advancement are within the occupation. The most obvious opportunities, or rewards, fall in the areas of increased wages and better — more attractive — types of runs. Some drivers eventually choose to move into dispatch or terminal manager positions. Still others become safety directors or driver trainers. And, of course, there are any number of drivers who happily stay behind the wheel for their entire career. The point is that, for the truly professional driver, the trucking industry is an open one with much to offer.

Applying for a Job

The job application process for tractor-trailer drivers generally involves an application form, reference checks, a job interview and some tests. There are few rules and regulations relating to applying for an intrastate hauling job. In applying for a job driving interstate, the basic process is about the same for all companies, because all of them must follow FMCSR job application and hiring requirements.

application form

Information supplied on the application form tells if you meet the basic job requirements. It can also give your potential employer a positive first impression, if it's done well. Your completed application should be neat, accurate and honest.

reference checks

When you go to apply for a job, always carry along the names and addresses of previous employers, as well as employment dates and supervisor's names. Potential employers will want to learn what previous employers have to say. For that reason, it's always best, if at all possible, to leave previous employers with a good impression. Even if you're unhappy with a job, it's in your own best interest to perform to the best of your ability right up to the end of the last day on that job.

the job interview

Next to the road test, the job interview is the most important part of the application process. Be prepared, know how to act and monitor your behavior, and end the interview well. This is your chance to sell yourself, make the best of it. Learn all you can about the company before going on your interview. Current employees, the local library, chamber of commerce and job service offices can all be good sources of information. Be prepared to ask intelligent questions. Know your own abilities and limitations. Make sure you have all the necessary paperwork along with you. That includes your CDL and related endorsements, certifications and letters of reference. Be ready. Get to the interview site a little early. (Visit the location a day or two ahead of time so you don't get lost.) Be well rested and neatly dressed. (Find out ahead of time, if possible, if you'll be asked to take a road test, so you can be appropriately dressed and prepared for it.) Sell yourself. Don't lie or brag, but present your qualifications and accomplishments honestly when asked. Relate your experience. Briefly state what you've done successfully in the past. Be positive. Don't dwell on bad experiences. Be responsive to questions asked. Answer questions, but don't volunteer unnecessary information. Try to keep the conversation focused on the job and why you're the best candidate for it.

The interviewer should volunteer information about the company, conditions and policies. If they don't, they expect you to ask. Don't wait until you get the job to find out it's not at all what you expected. Most of us have been blind sided at some point in our job history, but you can cut the chances of that happening by asking the right questions at an interview. Don't worry about being nervous. Most people are a little nervous at job interviews (including many interviewers). By yourself. Relax, and enjoy the experience. End the interview well. Thank interviewers for their time and interest. Express once more your interest in the job and the company, recapping important points. And finish by asking what will happen next; what procedure the company will follow in making a decision.

experience speaking

"The people who aren't professional, and, of course, we see almost as many of those as we do the people that are, generally, have certain immediate things at hand. How much money am I going to make? What do I have to do to get this money? Those are genuine concerns, of course, for anyone you are hiring, but they are, in many cases, not concerned about the long term objective of the industry or of their affect on the industry."

Hugh
Company Executive
Wisconsin

Tests

Be well rested and prepared to taken written and road tests. Be aware that a physical exam may be necessary.

written tests

Study key materials about regulations and safe driving practices. This would include, among others, any State driver's handbook and the FMCSR. Review information and procedures learned during your initial truck driver training.

road tests

Road tests are probably the single most important part of the application and hiring procedure. Be prepared for a test that could cover everything you've learned up to this point. This might include inspections, coupling and uncoupling, backing, parking and driving in traffic. Be dressed and ready for a road test. Have gloves and a log book along with you. Don't risk safety if you're not familiar with the vehicle. Ask about unfamiliar controls, parts or accessories. If confronted with a vehicle you're not qualified to operate, say so. This is no place to try to bluff your way into a job.

physical examination

You must have a current, valid certificate of medical examination showing that you meet minimum qualifications. If not, you will need to pass a physical for that purpose. Also, be prepared for the company to require its own physical for its own protection.

PROPER WORDS

Cracked, Bent, Broken, Missing, Securely Mounted

Springs: cracked, broken, missing leafs
Spring Hangers: Not cracked, bent, broken, missing, securely mounted

Rims: welds, cracks, dents
Tires: cuts, nicks, abrasions, bruises
Lug nuts: not cracked, bent, broken, missing, no bleeding, rust, securely mounted

Air hose: not rubbing, cracked, worn, leaking, securely mounted
Brake chamber: cracks, dents, leaks, securely mounted
Slack Adjuster: No more than one inch of travel, clevis pin has a cotterpin, not cracked, bent, broken, missing, mounted securely.
Brake lining: cracked, broken, missing pieces 1/4 inch minimum
Brake drum: cracked, broken, missing pieces, brake drum and lining are dry indicates oil seal is functioning properly

Air Hose - Brake Chamber - Slack Adjuster
Lining - Drum - Seal

the trucking profession review

1. What is the purpose for most laws governing the trucking industry?

2. Who is responsible for the public image of truck drivers?

3. Most conflicts with the public have to do with sharing the road: True or False?

4. During a conflict, what should you do if the other driver is at fault?

5. List three things a driver can do to help the motor carriers they work for keep business and get new business.

6. What do many fleet managers consider the number one requirement for a good driver?

7. Trucking is a career with limited opportunity beyond driving: True or False?

8. What types of tests are generally required as part of the application procedure for tractor-trailer operators?